Edwin James Houston

Outlines of forestry

The elementary principles underlying the science of forestry

Edwin James Houston

Outlines of forestry
The elementary principles underlying the science of forestry

ISBN/EAN: 9783337276041

Printed in Europe, USA, Canada, Australia, Japan

Cover: Foto ©berggeist007 / pixelio.de

More available books at **www.hansebooks.com**

Outlines of Forestry;

OR,

THE ELEMENTARY PRINCIPLES UNDERLYING THE SCIENCE OF FORESTRY.

BEING A SERIES OF PRIMERS OF FORESTRY.

BY

EDWIN J. HOUSTON, A.M.,

MEMBER OF THE PENNSYLVANIA FORESTRY ASSOCIATION, PROFESSOR OF PHYSICS IN THE FRANKLIN INSTITUTE OF THE STATE OF PENNSYLVANIA, PROFESSOR OF NATURAL PHILOSOPHY AND PHYSICAL GEOGRAPHY IN THE CENTRAL HIGH SCHOOL OF PHILADELPHIA, ETC., ETC.

PHILADELPHIA:

J. B. LIPPINCOTT COMPANY.

1893.

PREFACE.

When from any cause a necessity exists in any country for the removal of its forests from extended areas, unless care be taken as to the manner in which such removal is made, and some parts are left wooded, irreparable injuries will inevitably follow.

In the United States, where the enormous increase in population has resulted in the removal of the forests from extended areas, such intelligence and care have unfortunately, in most cases, not been exercised. The timber lands have generally been purchased at figures based almost entirely on the value of the standing wood. The trees have been cut down in a reckless manner, and fires, carelessly started, have often been left indifferently to burn themselves out. No attempts have been made to protect the soil that has been denuded of its natural protective covering by the axe or the fire. Before the forest has been made to yield its entire harvest, the greed of the

3

speculator has too often led him to abandon to the destructive action of the elements the area he has thus despoiled, in order to seek another, as yet unbroken, forest area.

Instead of carefully removing some of the trees from the forest, and leaving the area in such a condition as to enable it to produce a new growth, in the United States it has too frequently been the case that the virgin forest is thoughtlessly attacked, its best trees cut down in so careless a manner that the harvested crop amounts to, perhaps, but a third, or even less, of the total growth, and the remaining part abandoned to certain destruction by the elements.

The irreparable loss caused by such greed should be prevented by the enactment of judicious penal laws.

It is often very difficult to persuade the general public that evil results following any course of action, which do not come immediately, are not thereby prevented from coming eventually. Because the evil day draws not nigh quickly, there is a tendency to believe that it will never come at all.

An attempt has been made in the " Outlines of Forestry" to point out to the general public, in simple, non-technical language, the character of the effects, both on the general climate of a country and on the distribution of its rainfall, which inexorably follow the unsystematic removal of its forests.

It is only necessary to give the public some little insight into the effects produced by the destruction of the forest to arouse it to a conviction of the necessity for the existence of "Forestry Associations," for the enactment of laws regulating the manner in which the forests shall be removed, and for the setting aside of certain districts on which forests shall be perpetually maintained.

In order to enable the readers of this little book to carry their reading beyond the elementary principles which it discloses, appropriate extracts, taken from standard authors, and published by permission of the authors or publishers, have been added at the end of each primer. In all cases the exact title of the book has been given, as well as the names of its publishers.

For general aid in remembering the principles discussed in each primer, a concise review has been given at the close of the book in the form of a primer of primers.

The author has not hesitated to consult freely

all standard authorities in matters pertaining to the general subject of forestry.

There have been added to the book, in the shape of an appendix, lists of trees suitable for planting in different sections of the United States, as furnished by eminent authorities on the subject.

The author desires to express his thanks to the gentlemen who have responded to his circular letter of inquiry as to lists of trees suitable for replanting in various sections of the United States, and to his friend Professor Charles S. Dolley for revision of the manuscript of this book.

EDWIN J. HOUSTON.

CENTRAL HIGH SCHOOL, PHILADELPHIA, PA.,
January, 1893.

CONTENTS.

OUTLINES OF FORESTRY.

I. FORESTRY.

THE science of forestry not only treats of the care and preservation of those parts of the earth that are covered with trees, but also of the means best suited to the replanting of the areas from which the trees have been removed.

If nature is let alone she will cover any portion of the earth where vegetable life is possible with the particular kind of vegetation best fitted to grow under the existing conditions of soil, heat, light, and moisture.

Nature will, therefore, cover with forest all portions of the earth where forests are best fitted to exist.

It may then be asked, on what does the science of forestry rest? Why not leave nature alone? Such questions arise from a misunderstanding of what forestry endeavors to accomplish. Forestry does not aim to oppose nature, but simply to aid her. It endeavors to make use of the conditions naturally existing in any locality that are favorable to the continued growth of trees, and to oppose or hold in check conditions unfavorable to such growth.

The following considerations will suffice to show the necessity for the existence of Forestry Associations, and the enactment of strict laws for the care and preservation of the rapidly decreasing forest areas of the earth. The North Temperate zone, the cradle of the human race, possesses the densest civilized population. This zone was originally covered by extensive forests, and is still heavily wooded over extended areas.

Civilized man, however, cannot continue a dweller in the forest. It is true that in sparsely settled districts no necessity exists for the removal of the entire forest; but, as the density of population increases, the demands made on the forest increase. Such demands will, therefore, increase rather than decrease in the near future. Hence the necessity for the existence of Forestry Associations and the enactment of Forestry Laws.

The demands made on the forest by civilized man are either—

1. For the area on which the forest stands, or

2. For some of the products of the forest, such as wood for building purposes or fuel, or bark for the tannery.

The demands for the areas on which the forests stand are made either for the purpose of increasing the extent of the agricultural areas, or for the building of ordinary roads or railroads.

We may tabulate these demands as follows :

Encroachment on the forest for—

1. Area on which it stands.

   1. For land to be cultivated for ordinary farm products.

   2. For the construction of ordinary roads or railroads.

2. For its products.

   1. Fuel.
   2. Charcoal.
   3. Building purposes generally.
   4. Fences.
   5. Railroad ties.
   6. Telegraph poles.
   7. Mining purposes.
   8. Bark for tanneries.
   9. Turpentine, rosin.

When the forests are removed for the area on which the trees stand, the destruction is necessarily complete.

Where the removal is for agricultural purposes,

the destruction should not be complete. The principles of forestry teach that the truest economy will permit certain tracts to remain covered with trees; for the ultimate gain to the farmer from such a course will be greatly in excess of the sums paid for rental, or the interest-charges on the land that is not directly productive in ordinary farm products.

The areas required to be taken from the forests for agricultural purposes necessarily greatly exceed those required for the location of ordinary roads or railroads. The damage done indirectly to the forest, however, by the location of new roads, especially railroads, is often greater than by the location of new farming-lands, since the location of new roads greatly increases the liability to destructive fires, and also opens up extensive tracts of yet unmolested forest to the greed of the lumberman, or to the indifference of the railroad authorities or of the travelling public.

Besides this, an area taken for agricultural purposes is, to a certain extent, protected from the loss of its soil by a covering of vegetation. During the construction and operation of a road-bed, much of the adjoining land is often needlessly destroyed by being thoughtlessly left for floods to work irreparable damage by the removal of the soil.

It is for the purpose of regulating the necessary removal of the forest, and for pointing out the manner in which the products of the forest can be most advantageously harvested, that Forestry Associations and Forestry Laws are so imperatively demanded.

In order to intelligently protect forest areas, and thus aid rather than oppose nature in maintaining them, the principles underlying the growth of trees, the conditions of heat, light, and moisture, or, in general, the conditions of climate best suited to continue such growth, must be carefully studied. The natural influences or conditions which oppose the growth of trees must be ascertained, and, where possible, checked; the enemies of the forest recognized, and the best means taken to hold them in check. In other words, forestry must assume the position of an exact science, in order to call intelligently for the passage of laws intended for insuring the growth and reproduction of the earth's forests.

It is a mistaken idea that forestry endeavors to preserve intact the virgin forests of the earth. This is by no means the intent of intelligent forestry.

The wood and other products of the forest form an important part of the resources of a country. Man is as much entitled to the harvests of the woods as to the harvests of the fields. Forestry endeavors to point out the best ways in which forest crops may be harvested without detriment to the subsequent crops, and without causing the ultimate destruction of the forest.

Since it is manifestly impossible to preserve forests on all parts of the earth's surface where forests can naturally grow, it should be the duty and care of every community to set aside certain portions where forests shall be perpetually maintained; or, if already deforested, shall be replanted or reforested. Such preserves were originally maintained by the arbitrary will of the sovereign lord of the country, for the good of a few, in order to insure royal hunting-grounds. They should now be maintained, by the will of the people, for the good of the many.

The parts best suited for the perpetual maintenance of forests will necessarily vary in different regions.

In agricultural districts, certain areas should invariably be set aside on which trees shall be perpetually preserved; for true economy requires the maintenance of some timber on nearly all farm-land.

These areas will, perhaps, in the majority of cases be found as follows,—viz. :

1. On poor or thin soils where no other crops will thrive.

2. In damp places where no other crops will thrive.

3. On the borders of rivers or other streams.

4. On mountain slopes, hill-tops, or other elevations.

It can be shown, generally, that the areas which can be most profitably set aside for the maintenance of perpetual forests are situated, for the greater part, in the mountainous districts of the earth, the natural home of the forest.

As will hereafter be shown, the forests should especially be preserved on mountain slopes, for the following reasons :

1. Because the rain falls more frequently and in greater quantity in the mountainous districts of the earth than elsewhere. The cold mountain slopes, chilling the air, cause it to deposit its surplus moisture, in no matter from what direction the winds may come.

2. Because the principal rivers of the earth

are born in the mountainous districts, and, if the forests are removed, the rain which falls drains so rapidly from the earth's surface that the soil that took a long time to form by the gradual disintegration of the igneous rocks, and slowly accumulated its vegetable mould from the growth and subsequent decay of thousands of generations of plants, is lost to the highlands, only to become a source of damage to the lowlands.

3. Because the rapid drainage of the mountain slopes on the removal of their forests will result in dangerous floods during times of rain.

4. Because the failure of so great a part of the rain-water to sink into the ground and fill the reservoirs of the springs will cause such springs to more readily dry up shortly after the appearance of drought.

5. Because the mountain slopes, when deprived of their forests, become excessively hot during the day, and excessively cold during the night, and thus tend to sensibly alter the climate of the country.

6. Because such marked differences in temperature tend to increase the number and severity of destructive hail-storms.

7. Because the removal of the forests tends to

increase the liability of occurrence of early frosts in the neighboring agricultural districts.

8. Because the removal of the forests will be attended by marked changes in the relative quantity of moisture in the air at different times of the year.

B. E. Fernow, Chief of the Department of Forestry in the United States Department of Agriculture, in a paper entitled " What is Forestry?" * says, on page 15 :

" Forestry in a wooded country means harvesting the wood crop in such a manner that the forest will produce itself in the same, if not in superior, composition of kinds. Reproduction, then, is the aim of the forest manager, and the difference between the work of the lumberman and that of the forester consists mainly in this : that the forester cuts his trees with a view of securing valuable reproduction, while the lumberman cuts without this view, or at least without the knowledge as to how this reproduction can be secured and directed at will. The efficient forest manager requires no tool other than the axe and the saw,—the planing-tools being only needed to correct his mistakes,—but he sees them differently from the lumberman."

---

* Reprinted, by permission, from "What is Forestry?" by B. E. Fernow, Washington, Government Printing-Office, 1891.

The true position which forestry takes in the United States is thus forcibly expressed in the Annual Report of the New York Forest Commission, for the year ending December 31, 1890,* on page 91, as follows:

"A misunderstanding has prevailed to some extent with regard to the attitude of forestry towards the lumber interests of private owners. It is, however, generally misunderstood, now, that the true interests of the lumbermen are not incompatible with forest preservation, and it has been declared to be one of the objects of the forestry movement in this country 'To harmonize the interests of the lumberman and the forester, and to devise for the lumbering interest such protection as is not given at the cost of the forests.' Forestry is not opposed to having trees cut down in the proper way. They must be cut to supply the world with timber. They furnish the material for shelter to mankind, and contribute to render the houses of men comfortable and beautiful by providing fuel and decorations. It is needless to point out here the manifold purposes for which wood is needed, and how largely it enters into our industries and arts, contributes to our convenience and pleasure, and becomes a necessity of our daily lives. Civilization could hardly exist without it. It is from trees, and from trees only, that our needs for wood are supplied

* Reprinted, by permission, from Annual Report of the New York Forest Commission, for the year ending December 31, 1890. Albany: James B. Lyon, State Printer, 1891. Pp. 317.

through the timber-dealer and lumberman. It is not the exercise of their vocation, but their frequent abuse of it, that calls for criticism,—a distinction that has not always been made by the critics. Estimates show that thirty billion feet are required annually in this country for building and manufacturing purposes alone, leaving the fuel question out of consideration. It is the unwise, improvident, stupid method, or want of method, by which the cutting has heretofore too often been done, that is deplored. Under the old practice the forests have rapidly disappeared, and, if it continues, in a few years none will be left. The lumberman will have ruined his own business, as there will be no forests to furnish him with his stock in trade. It is the purpose of forestry to point out to the lumberman the true methods of exercising his own profession, which will provide him material for the future, as well as the present, by maintaining permanent forests through a succession of crops."

II. CONDITIONS NECESSARY FOR THE GROWTH OF PLANTS.

EXTENDED scientific research has established the fact that all forms of life, whether of the animal or the plant, can, at their earliest stages, be traced to a minute germ-cell filled with a more or less transparent substance called protoplasm, and containing a dark, opaque spot called the nucleus. Unless this germ-cell exists, plant or animal life is impossible.

Although cases exist where there has been no apparent evidence of the presence of a germ or seed, yet such germ or seed must have existed, and was derived from a plant of exactly the same character as that which such seed will produce when called into active and matured growth.

Under peculiar circumstances, plant or animal germs possess wonderful vitality, and may remain in a dormant state for a very long time, only beginning to grow when exposed to the conditions necessary for growth.

In order that any form of plant life may exist

on the earth, the following conditions are necessary :

1. The germ or seed from which the plant grows.

The germ or seed in all cases comes directly from a plant similar to that which is produced when the seed sprouts or germinates, and attains its full growth.

2. The cradle where the plant is born.

The plant's cradle is the soil. In the soil the plant spreads its roots, and from it obtains, in great part, the materials necessary for nourishment and growth.

Cases exist where the plant finds its cradle in the air or in the water. These need not, however, be considered in this connection.

3. The sunshine and the heatshine, which awaken the sleeping germ and call it into activity; or, in other words, the light and heat which are so essential to a plant's growth.

4. The nourishment, or the food which the plant takes into its structure and assimilates, and thus causes it to become a part of itself.

The processes by which a plant causes different materials taken from the soil in which it grows, or from the air around it, to become a part of itself, is called assimilation. At the commencement of its life the plant gets its nourishment from the protoplasm surrounding the nucleus. It is not long, however, before it exhausts this stock of food, and it must then get all its nourishment either from the soil or from the atmosphere; or, in other words, from outside the seed.

This latter nourishment of the plant comes from a variety of materials, derived either from the air or from the ground, the most important of which are as follows :

1. Moisture.

This moisture is mainly taken up by the roots of the plant from the soil; but it is, in some cases, absorbed directly from the air by the leaves.

2. Carbonic acid.

Carbonic acid is a gaseous substance, formed of carbon combined with an invisible gas called oxygen. The carbonic acid is absorbed by the leaves of the plant, and, in the presence of sunshine, is broken up into carbon and oxygen. The oxygen is given off from the surfaces of the leaves, and the carbon is retained by the plant to form its woody fibre. In the case of large vegetable forms like forest-trees, the amount of carbonic acid taken from the air and converted into woody fibre must be very great. A certain amount of oxygen, however, must be present in the air, to permit the continued growth of the plant; for most plants will not grow in an atmosphere of pure carbonic acid gas.

The hydrogen needed for the plant's growth is derived from the decomposition of the water associated with the carbonic acid; the result is that the plant retains the carbon and the hydrogen, and throws out the oxygen into the atmosphere.

3. Mineral matters taken from the soil.

The tissues of the plant contain various kinds of mineral substances which are taken directly from the soil. For the proper growth of the plant, the soil must contain these particular mineral substances in the condition or state in which they can be readily taken up or assimilated by the plant.

The above conditions—viz., the germ, the cradle, the sun's light and heat, and some form of solid and liquid food—are not of equal importance to the growth of the plant.

The presence of the germ or seed is, of course, of the greatest importance, since without it no plant can grow.

The sunshine and the heat may, perhaps, be considered as next in importance to the growth of the plant. Heat and light are to be found in practically all parts of the earth. They differ, however, in amount, in different regions of the earth, and such differences cause the differences that are noticed in the plants that grow in different regions.

Even in the same region, differences in the light and heat cause differences in the plant's growth, as may be noticed in almost any forest region.

The plants that form the undergrowth of those portions of the forest where the light is more thoroughly shut out are markedly different from those appearing in the clearings, where the light and heat have full access to the soil.

The nourishment of the plant comes next in importance. The quantity of carbonic acid found in the air is practically the same in all parts of the earth. The quantity of moisture in the air differs very greatly in different parts of the earth, and on this difference, together with the difference of temperature, depends the differences observed in the plants of various regions.

The liquid nourishment of the plant in the shape of water is of so great importance to the growth of the plant, that the character of the rain-fall in any country will, to a marked extent, determine the character of the flora of that country.

The soil is, perhaps, the least important of the conditions required for plant growth.

This statement is at first thought so much at variance with the generally received opinion as to need an explanation.

Where a particular kind of plant is to be grown, the character of the soil, probably, stands next in importance to the presence of the germ or seed; for each plant thrives best in a particular kind of soil. The variety of plants that exist, however, is so great that, given almost any kind of soil, together with certain conditions of heat, light, and moisture, such soil will be found to be best suited for the growth of some particular kind of plant. In other words, if the proper conditions of moisture, heat, and light are present, and the germ is present, vegetation will appear in almost any region.

Nature has generously scattered the germs of various forms of plant life nearly all over the earth's surface. Therefore, if unmolested by man, she will, in most cases, maintain on such surfaces the kind of plant forms or plant growths best suited to grow naturally.

There, therefore, will be found in every section of country a plant growth or plant life peculiar to, or naturally belonging to, such a section of country. Each section of country possesses, so to speak, a nationality in its plants, or, in other words, there lives in each section of country a particular nation of plants. Such a nation of plants, or the plants peculiar to a particular section of country, is called its flora.

Since heat, light, and moisture are, next to the presence of the plant germ, the most important things for plant growth, there will necessarily exist in different parts of the earth a flora that will vary according to the differences that exist in the distribution of heat, light, and moisture over such part of the earth's surface.

The heat, light, and moisture are greater in amount at the equator than at any other portion of the earth's surface. Therefore, the vegetation is more luxuriant, or possesses a greater diversity of forms, here than at any other part of the surface. As we pass from the equator towards the poles, the decrease in the heat, light, and moisture causes a corresponding decrease in the variety and luxuriance of vegetable life.

In passing from the base to the summit of a

high tropical mountain, the same differences in the variety and luxuriance of plant life are noticed that are seen in going from the equator to the poles. This is due mainly to the distribution of the heat and moisture.

The planting of a germ or seed in any soil will not result in its continued growth, unless the conditions of heat, light, and moisture are practically the same as those in which the plant from which such germ or seed was derived required for its existence.

Trees planted in a particular locality may, therefore, fail to grow in such locality, from want of the proper conditions of heat, light, and moisture.

In all regions where forests can grow naturally, wherever practicable, they should be permitted to grow, since, as will be shown, the continued existence of forests on certain portions of the earth is necessary for insuring that balance of nature on which the comfortable existence of man depends.

Guyot, on page 188 in his "Earth and Man," *

---

* "The Earth and Man," lecture on Comparative Physical Geography in its Relation to the History of Mankind, by Arnold Guyot, Professor of Physical Geography and History at Neufchâtel, Switzerland. Boston : Gould, Kendall & Lincoln, 1849.

thus refers to the conditions favorable to luxuriant
vegetable growth:

"The warm and the moist—these are the most favorable
conditions for the production of an exuberant vegetation.
Now, the vegetable covering is nowhere so general, the vegetation so predominant, as in the two Americas. Behold,
under the same parallel where Africa presents only parched
table-lands, those boundless virgin forests of the basin of the
Amazon, those selvas, almost unbroken over a length of more
than fifteen hundred miles, forming the most gigantic wilder-
ness of this kind that exists in any continent. And what
vigor, what luxuriance of vegetation! The palm-trees, with
their slender forms, calling to mind that of America itself,
boldly uplifted their heads one hundred and fifty or two hundred feet above the ground, and domineer over all the other
trees of these wilds, by their height, by their number, and by
the majesty of their foliage. Innumerable shrubs and trees
of smaller height fill up the space that separates their trunks;
climbing plants, woody-stemmed, twining lianos, infinitely
varied, surround them both with their flexible branches, dis-
play their own flowers upon the foliage, and combine them in
a solid mass of vegetation, impenetrable to man, which the
axe alone can break through with success. On the bosom of
their peaceful waters swims the Victoria, the elegant rival of
the Rafflesia, that odorous and gigantic water-lily, whose white
and rosy corolla, fifteen inches in diameter, rises with a daz-
zling brilliancy from the midst of a train of immense leaves,
softly spread upon the waves, a single one covering a space of
six feet in width. The rivers, rolling their tranquil waters

under verdurous domes, in the bosom of these vast wilds, are the only paths that nature has opened to the scattered inhabitants of these rich solitudes. Elsewhere, in Mexico and Yucatan, an invading vegetation permits not even the works of man to exist; and the monuments of a civilization comparatively ancient, which the antiquary goes to investigate with care, are soon changed into a mountain of verdure, or demolished, stone after stone, by the plants piercing into their chinks, pushing aside with vigor, and breaking with irresistible force, all the obstacles that oppose their rapid growth."

The author, in his " Elements of Physical Geography," * page 119, thus refers to the growth of living matter :

" All life, whether vegetable or animal, consists of various groupings of cells, or approximately spherical masses, consisting of a peculiar form of a jelly-like matter called *protoplasm*, composed of various complex combinations of carbon, hydrogen, oxygen, and sulphur, called *proteids*. At its beginning all life consists of a minute germ-cell, filled with more or less transparent protoplasm, and containing a darker opaque spot called the *nucleus*. Examined by a sufficiently powerful glass, all living protoplasm is seen to be in constant motion, currents passing through the different parts in somewhat definite directions.

---

* Reprinted, by permission, from " The Elements of Physical Geography, for the Use of Schools, Academies, and Colleges." By Edwin J. Houston, A.M. Eldredge & Brother, No. 17 North Seventh Street, Philadelphia. 1891. Pp. 272.

"As the germ-cell develops in all the higher forms of life, it multiplies, and various organs appear, peculiar to the form of life from which the germ-cell was derived. All living bodies contain organs, and living matter is therefore sometimes called *organic matter*, to distinguish it from non-living or *inorganic matter*.

"Science has not yet disclosed the nature of the change whereby non-living matter is converted into living protoplasm. To produce living matter, the intervention of already living matter is, so far as is known, absolutely necessary."

Concerning the influence of climate on plant growth, Élisée Reclus, in his work, "The Ocean,"* on page 361, says:

"Each plant has its special domain, determined not only by the nature of the soil, but also by the various conditions of climate, temperature, light, moisture, the direction and force of winds, and of oceanic currents. During the course of ages the extent of this domain changes incessantly, according to the modifications which are produced in the world of air, and the limits of the region inhabited by the various species are dovetailed into one another in the most complicated manner. The flora indicates the climate; but what is the climate itself, in the apparently confused mixture of phenomena which compose it? The preponderating influence

* Reprinted, by permission, from "The Ocean, Atmosphere, and Life," by Élisée Reclus. New York: Harper & Brothers, Publishers, Franklin Square. 1874. Pp. 534.

is naturally that of temperature; nevertheless, we must not think, as many botanists did till very recently, that the limits of the zone of vegetation of each plant are marked on the continents by the insinuosities of the isothermal lines. In fact, as Charles Martins and Alphonse de Candolle remark, each plant requires for its germination and development a certain amount of temperature, differing according to the species. With some, life resumes its activity after the sleep of winter, when the thermometer marks three or five degrees above the freezing-point; others need a heat of eighteen, twenty, and even twenty-five and thirty-five degrees, before taking the first step in their career of the year. Each species has, so to say, its particular thermometer, the zero of which corresponds to the degrees of temperature when the vegetating force awakens its germs. It is, therefore, impossible to indicate by such general climatal lines the limits of habitation for such or such species, since each one of them has for the commencement of its vital period a different starting-point."

III. THE WIDE DISTRIBUTION OF PLANT GERMS.

In order that seed-time and harvest shall not fail on the earth, nature has distributed the seeds or germs of plant life with a liberal hand over all parts of the surface. Even amid the burning sands of the deserts and the eternal snows of the polar regions, myriad forms of plant germs exist.

Nature has provided numerous ways for insuring the thorough scattering of these plant germs or seeds.

Many forms of seeds are provided with delicate hair-like projections, or with wings, by means of which they are carried by the winds to great distances. Others are provided with projecting hooks or bristles, by means of which they catch in the fur of animals, or in the plumage of birds, and are thus carried into distant regions.

Perhaps one of the most important of the means provided by nature for the distribution of plant germs, is that the seeds, which are swallowed whole by birds or other animals, subsequently pass

out of the animal uninjured by the process of
digestion. By such means seeds are carried to
distant parts of the earth.

Man, either purposely or accidentally, scatters
plants, germs, or seeds in far-distant countries.
Not unfrequently some of the plants thus brought
from one country to another find the conditions
of soil and climate in their new home so favorable
to growth as to completely drive out and exterminate domestic species.

Besides the means just mentioned for the scattering of germs or seeds of plants, there are possibly
others that have not yet been recognized.

The germs or seeds of plants possess a singular
vitality under certain conditions. The grains of
corn or wheat found in the Egyptian mummies, in
many cases, grew and bore fruit notwithstanding
their centuries of rest. Such instances of the
preservation of vitality are, perhaps, less wonderful when viewed in the light of the exceedingly
dry climate in which the mummies were preserved.
More curious instances are found in which the
germs existed for a long time in the presence of
an abundance of moisture, and did not grow as
long as the heat and light alone were absent.

The truth of the above statements is denied by

c

some, but that such experiments succeeded in
some cases is undoubtedly true.

For example, in densely-wooded countries, where
the ground is thickly covered with trees, the light
and heat of the sun are so thoroughly expended
in maintaining such growth that no other forms
of plant life occur. Let, however, some of the
trees be removed, so that the light and heat of
the sun may reach the ground, and the seeds that
were there, possibly during the centuries that the
forest covered it, at once spring into active life.
Here all the conditions except sufficient light and
heat were present, and yet the germs slumbered.

That the Sahara desert was once, in certain
portions, if not in all parts, well watered, is
attested by the presence of the wadys, or deserted
river-valleys. That the soil of the desert contains
a liberal supply of numerous plant germs, is shown
by the fact that, on the successful sinking of an
artesian well, the appearance of the water is invariably attended by the appearance of a flora
often containing peculiar species of plants. Here
the light, heat, and soil were all present, and yet
the germs slumbered for want of moisture.

The boring of artesian springs, or the digging
of cellars or other excavations, by bringing to the

sun's light and heat soil which has been deprived of such sunlight and heat for unknown ages, in many—indeed, in most—cases is followed by the appearance of vegetation often containing species quite strange to that particular section of country.

A curious story of such a case is told, which, if true, shows in a striking manner the wonderful vitality of certain seeds. In a given section of country, let us say in England, a farmer commenced to dig a well. This act, so common in an agricultural district, attracted no particular attention until the depth of the still dry hole far exceeded that of most wells in the locality. The neighbors then began to speculate as to whether the farmer would eventually strike water, but when he continued without success, many of his neighbors began to quietly laugh at him. The farmer, however, persisted, and at last his persistency or stubbornness, whichever it may have been, was rewarded.

After having dug through a considerable deposit of sand, which appeared very much like the sand of an ancient sea-beach, a water-logged stratum was reached from which gushed forth a copious supply of excellent water. So pleased was the farmer with the result of his labors, that he arranged the sand and other materials brought up from below in a form of a garden-plot around the mouth of the well. Strange plants soon appeared, and among them a tree, which, when sufficiently matured, proved to be an ancient form of beech-plum. Now, since it is universally recognized that no form of plant life appears without the presence of a germ similar to that which the plant will itself produce, the germ of this ancient beech-plum was presumably preserved in the deep-seated strata for untold centuries, awaiting to be called into life by the genial warmth and light of the sun.

The virgin soil of the prairies, where turned up by the plough of the settler, generally produces a vegetation different from that of the undisturbed soil. Even the tracks of the settlers' wagons disturb the soil sufficiently to be afterwards marked by a growth of plants quite distinct from those which cover the undisturbed portions. The seeds must have lain a long time below the surface, only springing into active life on exposure to the light and heat of the sun.

The burning of a pine forest in the North Temperate Zone is almost invariably followed by a growth of scrub oak. What was the origin of the germs of these oaks? They presumably existed in the ground, and, in some as yet unexplained manner, were aroused into active life by the presence of the fire.

In some parts of the United States the burning over of a region is almost invariably followed by a growth of what is very appropriately called fireweed, the seeds of which appear to have been called into active life in some as yet unexplained manner, either by something added to the soil by the heat, or possibly by the heat of the fire itself.

Cases are on record where earthquakes have brought up to the surface, soil which had probably been buried for ages, but which, on exposure to the light and heat of the sun, gave birth to strange forms of plant life.

Possibly, in some of the cases mentioned, the germs of plant life have been carried to the localities by one or another of the agencies already mentioned. In other cases, however, the germs appear to have existed in the soil, waiting to be called into life by the sun's light and heat.

Nature, therefore, has taken care that the earth shall be covered with a vegetable carpet wherever man does not oppose her action. If left to work out her own course, she will cover the earth with a dress of such vegetable forms as are best suited

4

to exist there naturally. If interfered with, to even a comparatively trifling degree, such changes may be produced in the soil, climate, or other conditions as will bring wide-spread destruction to widely separated sections of the country.

Speaking of the vitality of seeds, Marsh, in his work entitled " The Earth as Modified by Human Action," * page 295, says:

" When newly-cleared ground is burnt over in the United States, the ashes are hardly cold before they are covered with a crop of fire-weed, *Senecio hieracifolius*, a tall, herbaceous plant, very seldom seen growing under other circumstances, and often not to be found for a distance of many miles from the clearing. Its seeds, whether the fruit of an ancient vegetation or newly sown by winds or birds, require either a quickening by a heat which raises to a high point the temperature of the stratum where they lie buried, or a special pabulum furnished only by the combustion of the vegetable remains that cover the ground in the woods.

" Earth brought up from wells or other excavations soon produces a harvest of plants often very unlike those of local flora, and Hayden informs us that on our great Western desert plains, wherever the earth is broken up, the wild sunflower (Helianthus) and others of the taller-growing plants, though

---

* Reprinted, by permission, from " The Earth as Modified by Human Action," by George P. Marsh. New York: Scribner, Armstrong & Co., No. 654 Broadway, 1874. Pp. 656.

previously unknown in the vicinity, at once spring up, almost as if spontaneous generation had taken place."

The wonderful vitality of certain seeds is thus referred to by Lindley in his "Botany," * on page 358:

"The action of seeds is confined to that phenomenon which occurs when the embryo which the seed contains is first called into life, and which is named germination.

"If seeds are sown as soon as they are gathered, they generally vegetate, at the latest, in the ensuing spring; but, if they are dried first, it often happens that they will lie a whole year or more in the ground without altering. This character varies extremely in different species. The power of preserving their vitality is also variable: some will retain their germinating powers many years, in any latitude, and under almost any circumstances. Melon-seeds have been known to grow when forty-one years old, maize thirty years, rye forty years, the sensitive plant sixty years, kidney-beans one hundred years. Clover will come up from soil newly brought to the surface of the earth, in places in which no clover had been previously known to grow in the memory of man, and I have at this moment three plants of raspberries before me, which have been raised in the garden of the Horticultural Society from seeds taken from the stomach of a man whose skeleton was found

---

* "An Introduction to Botany," by John Lindley, Ph.D. London: Longman, Orme, Brown, Green, and Longmans. Third Edition, 1839. Pp. 594.

thirty feet below the surface of the earth, at the bottom of a barrow which was opened near Dorchester. He had been buried with some coins of the Emperor Hadrian, and it is therefore probable that the seeds were sixteen or seventeen hundred years old."

. " It has already been seen that under certain circumstances, the vitality of seeds may be preserved for a very considerable length of time; but it is difficult to say what are the exact conditions under which this is effected. We learn from experiment that seeds will not germinate if placed *in vacuo*, or in an atmosphere of hydrogen, nitrogen, or carbonic acid; but no such conditions exist in nature, and, therefore, it cannot be they which have occasionally preserved vegetable vitality in the embryo plant for many years. Perhaps the following remarks, in a work lately published by the Society for the Diffusion of Useful Knowledge, may throw some light on the subject:

" It may, upon the whole, be inferred from the duration of seeds buried in the earth, and from other circumstances that the principal conditions are, 1, uniform temperature; 2, moderate dryness; 3, exclusion of light; and it will be found that the success with which seeds are transported from foreign countries, in a living state, is in proportion to the care and skill with which these conditions are preserved. For example, seeds brought from India, round the Cape of Good Hope, rarely vegetate freely: in this case the double exposure to the heat of the equator, and the subsequent arrival of the seeds in cold latitudes, are probably the causes of their death; for seeds brought overland from India, and therefore not exposed to such fluctuations of temperature, generally succeed. Others, again, which cannot be conveyed with certainty if exposed to the air, will travel in safety for many months if buried in clay rammed hard in boxes: in this manner only can the seeds of the mango be brought alive from the West Indies; and it was thus the principal part of the Araucania pines, now in England, were transported from Chile. It may therefore be well worth consideration, whether by some artificial contrivance, in which these principles shall be kept in view, it may not be possible to reduce to something like certainty the preservation of seeds in long voyages,—such, for instance, as by surrounding them with many layers of non-conducting matter, as case over case of wood, or by ramming every other space, in such cases, with clay in a dry state."

4*

IV. CONDITIONS NECESSARY FOR THE GROWTH OF TREES.

If a soil exists in any locality, and certain conditions of light, heat, and moisture are present, the character of the vegetation that naturally grows in such a region will depend more on the peculiarities of the distribution of the heat, light, and moisture, than on the character of the soil itself.

If moisture be entirely absent, or if it exists in such a form as ice or snow, in which it cannot be readily appropriated by plants, then that region must become a desert.

Deserts occur either in dry, arid regions, or in the regions of perpetual snow of the polar zones, or on the higher mountain slopes.

If the rainfall is absent during certain seasons of the year, but occurs during the rest of the year,— that is, if one part of the year is dry and the rest is wet,—the vegetable forms, which die or disappear at the beginning of the dry season, reappear at the beginning of the wet season.   Areas of the earth possessing this character of vegetation are called steppe regions.

When the rainfall is not very great in amount, but is fairly well distributed throughout the year, so that the rain is never absent for a very long time, regions called meadows or prairies occur.

If there is an abundance of moisture at nearly all times throughout the year, so that such moisture is absent for no very long time, then the country may be covered by trees. Such areas are called forests.

Forests cannot exist in the temperate zones of the earth in localities where, during the time of the trees' active growth, a very long interval exists during which no rain falls. While the active growth of the trees is temporarily suspended, as during the winter, this necessity for liquid nourishment, of course, no longer exists.

The reason forests cannot grow except where moisture is present during nearly all the time the plants are growing, will be easily understood from the following considerations:

Suppose a soil exists in any section of country, and such soil contains germs of practically all species of plant life. When such a soil is submitted to the action of light, heat, and moisture, most of these germs will be called into active growth, and various forms of plant life will begin their existence.

Suppose this particular section be a region where, for several months of the year, no rain falls, and whose soil, as is generally the case, rapidly becomes dry. During the dry season all forms of plant life will die from want of proper nourishment

On the reappearance of the wet season, only those forms of plant life will appear that have been able, during the brief time of the wet season, to reach their full maturity and produce their fruit or seeds, and so supply the germs necessary for a new growth. Such forms as trees, which, as is well known, require many years to mature their seed or fruit, will necessarily be unable to continue to grow naturally in such a region of country.

Of course, it might easily happen that during the first wet season all the germs might not have been called into active life by the combined influence of the light and heat, so that on the next wet season such forms might again spring up naturally.

But their continued existence, under these circumstances, would be impossible from the absence of the new germs.

In any section of country where the rainfall is limited to certain periods of the year, only those plants can continue to grow that, during the time the rain continues and water is supplied to them, or during the time the plant is actually growing, can reach their maturity and develop their seeds, and thus supply new germs that shall be ready for the appearance of the next rainy season.

For the growth of forests, both a certain depth of soil and, in general, a certain character of soil are necessary. This soil was slowly formed by the decomposition of the hard, igneous rocks that originally formed the entire crust of the earth, and contains a quantity of vegetable mould or humus derived from many successive generations of plant.

In the beginning, when the rocks' bare surfaces emerged from the universal oceans, forests could not grow even where the proper conditions of light, heat, and moisture were present, until such soil had been prepared for them.

Extensive forests can exist naturally only in regions where suitable soil exists and where the rainfall during the time of growth is maintained with a certain approach towards regularity, so that the trees are then properly and continually supplied with liquid nourishment.

It is in the temperate regions of the earth and in some parts of the tropics that the great forest areas are to be found, since it is in these regions that the rain may fall at almost any time of day, and on almost any day of the year.

There are, however, certain regions in the tropics where forests exist, although there are comparatively extended periods during the growth of the trees when rain does not fall. Here, however, the air is very moist, and heavy dews take the place of rain, or the rich vegetable humus absorbs the vapor directly from the air; or, in some cases, though growth is not actually suspended, it is at least so markedly retarded that the decreased nourishment of the trees is less injurious.

It is especially on the sides of mountains, where rain may fall in no matter from what direction the wind comes, or on the side of an island or continent that receives the prevalent wind, that forests are to be found in nearly all portions of the earth's surface, provided the heat is sufficiently great and a suitable soil is present. These conditions of soil and temperature exist on nearly all mountain slopes outside the polar regions.

The mountains may, therefore, be regarded as the natural home of the forests. The mountains are also the natural birthplaces of the rivers. The preservation of the forests on the sides of mountains is necessary to insure such a drainage of the rainfall as will best preserve the uniform flow of the rivers and best prevent them from overflowing their banks in times of rain, or becoming too shallow in times of drought.

The preservation of the forests is necessary in certain portions of the lowlands to protect the crops from the direct action of winds that are either too hot or too cold.

When it is necessary to cut down the forests for the sake of their timber, the areas on which they grew in all cases should be replanted, so that such areas may be able to yield continually successive crops of timber so necessary for man's needs.

As to the necessity for an abundance of water well distributed throughout the year in order to insure the growth of trees, Guyot, in his " Earth and Man," * says, on page 189 :

---

* " The Earth and Man : Lectures on Comparative Physical Geography in its Relation to the History of Mankind," by Arnold Guyot. Boston: Gould, Kendall & Lincoln, 59 Washington Street, 1849.

"North America, in spite of its more continental climate, shares no less in this character of the New World. The beauty and the extent of the vast forests that cover its soil, the variety of the arborescent species composing them, the strong and lofty size of the trees which grow there, all these are too well known for me to stop and describe them. It is because to a more abundant irrigation this continent adds a soil slightly mountainous, almost everywhere fertile, securing it always an equal moisture, a more abundant harvest of all the vegetables useful to man."

Concerning the growth of trees on mountain slopes, Élisée Reclus, in his work, " The Ocean," * says on page 383:

"The stages of vegetation have been studied with care on the slopes of many other mountains of temperate Europe, especially on the sides of the Ventoux, by M. Charles Martins; but it is in the Alps, above all, that the most celebrated botanists of our country have made their comparative researches on the floras of the various altitudes. The limits of these floras vary, so far as we can understand, according to the form, exposure, and height of the mountains, the nature of the rocks, the moisture of the soil, and abundance of snow, and the meteorological conditions of the surrounding atmosphere. It is, therefore, impossible to give the precise figures on the whole of the Alpine masses, and the averages obtained by

* Reprinted, by permission, from "The Ocean, Atmosphere, and Life," by Élisée Reclus. New York: Harper & Brothers, Publishers, Franklin Square, 1874. Pp. 534.

savants have only a very general value. Without taking account of the upper limits of cultivation, which varies singularly in the high valleys in proportion to the industry, intelligence, and social condition of the inhabitants, we may say that the vegetation of the plain hardly exceeds three thousand feet; above this height the slopes where man has not violently interfered to change the productions of the soil are naturally covered by vast forests. Still, the great trees gradually diminish in height in proportion as we rise into a zone where the air is rarer and colder; their wood becomes harder and more knotted; and the hardy kinds, which venture not far from the region of the snows, end by creeping on the ground, as if to seek shelter between the stones. To the north of Switzerland, the beech does not exceed the height of four thousand feet, and the spruce-fir stops at six thousand feet. In the group of Monte Rosa, the same forest growth, which approaches most nearly to the zone of perpetual snow, ascends as far as six thousand two hundred feet on the northern slope; while on the opposite side the larch, still hardier, attains its upper limit at seven thousand two hundred feet. Higher still we only find the fantastically twisted trunks of a few *mugho* pines, rhododendrons, willow-herbs, and juniper-trees; then all vegetation becomes more stunted, and is attached to the ground in order to escape the icy winds, and to allow of its being covered in winter with a protecting layer of snow up to the very edges of the glacier and the white surface of the snows."

V. THE FORMATION OF SOIL.

THE soil, in which the plant grows and which forms its cradle, is composed chiefly of mineral matters derived from the originally crystalline rocks which were formed by the gradual cooling of the earth's crust.   The soil, however, also contains a small quantity of vegetable mould or humus, obtained by the growth and subsequent decay of successive generations of plants.

Before soil can be formed, the hard crystalline rocks must be broken up, or, as it is technically called, disintegrated.

This disintegration is effected to some slight degree by the roots of plants, but it is mainly brought about by the action, in one way or another, of water.

Soils are divided by Gray into gravelly, sandy, clayey, calcareous, loamy, and peaty.

Gravelly soils are such as have coarse pebbles or fragments of quartz, lime, or feldspar spread through more finely divided mineral matter.

Sandy soils are usually formed of fine particles

of quartz, associated with feldspar. Such soils generally contain some compound of iron.

Clayey soils are formed from the decomposition of feldspathic rocks. They are impervious to water and harden on drying.

Calcareous soils contain carbonate of lime in large amounts.

Peaty soils are such as contain a large proportion of partially decayed vegetable matter.

The soil is sometimes found resting in place, directly on top of the rocks from which it was derived. In such cases its various mineral ingredients can be traced directly to the decomposition of the underlying rocks.

A section of such rock and soil will show the rock gradually passing from the loose soil into the hard and unchanged rock.

In other cases the soil is found at a considerable distance from the locality in which it was originally formed, or in which the plants grew that produced its vegetable mould.

The soil is carried from the rocks from which its mineral ingredients were derived either by the action of the winds or by the waters, though mainly by the latter, to distances which often reach thousands of miles.

Where the soil remains directly on the surface of the rocks from which it was derived, it is interesting to trace the gradual changes that occur in passing downward from the loose soil to the hard, unchanged rock below. On top is the loose soil, with the admixture of vegetable humus so necessary to the growth of the higher forms of vegetable life. Under this is thoroughly broken-up rock, which contains less vegetable matter. Under this is coarser and less broken-up rock. Under this the rock is intact and merely softened by the agencies effecting the disintegration. Finally, lying under all, is the still untouched virgin rock.

The principal agencies causing the disintegration of rocks are:

1. The expansive force which sprouting or growing vegetation exerts on the rock.

2. The alternate expansions and contractions that attend the freezing and thawing of the water which flows into the crevices between the rocks, or sinks into their porous structure.

3. The erosion or cutting power of water charged with suspended matter or sediment.

4. The erosion or cutting of glaciers or masses . of moving ice.

5. The solvent power of water, especially when

aided by the chemical action of such gases as oxygen and carbonic acid gas dissolved in the water.

During the vigorous growth of any form of plant life the increase in the length and diameter of the roots will break up or disintegrate even the hardest of rocks. This action is especially effective from the fact that in many cases the roots extend for the greater part through crevices or cracks where the quantity of moisture is greater than elsewhere.

The effects of alternate expansion and contraction are limited to climates where the temperature occasionally falls below the freezing-point of water. The water sinking into the porous rocks, and filling the crevices and cracks between them, expands on freezing and breaks the rock into fragments. These fragments are afterwards broken into smaller fragments, until the pieces are sufficiently small to be carried by the winds and waters to distant localities.

· The ability of running water charged with suspended mineral matter to cut or wear away hard rocks is very great. The moving water carries mechanically suspended in it minute fragments of such hard minerals as angular fragments of quartz or pebbles. These act as planes, cutters, or chisels that gnaw, cut, or wear away even the hardest rock. This process is technically called erosion, and is of great aid in the formation of soil.

There collects on the sides of mountains above the limit of perpetual snow an immense accumulation of snow, which, through gradual pressure, is converted into hard ice, and forms masses called glaciers. The glaciers slowly move or slip down the sides of the mountain. They receive the drainage of snow from the slopes of the valleys through which they move, just as rivers receive the drainage of liquid water. The glaciers carry with them considerable mineral matter, both in the shape of small rocks and large boulders. As the mass moves down the mountains, this mineral matter is pressed against the sides of the valleys, or along the bottom of the bed through which it is moving, and cuts, grooves, or grinds the hard rocks, and thus aids in the production of soil.

From its great solvent power, water is able to finally sink into what were originally impervious rocks, by gradually dissolving out the soluble portions of the rocks. In this way the rock is rendered rotten by the removal of the materials which formerly acted as a cement to bind its different ingredients together. When charged with either oxygen or cabonic acid gas, derived generally directly from the air, the chemical action of these dissolved gases greatly aids the water in breaking up, and thus rendering partly porous, even the hardest of the igneous rocks.

Let us take, for example, some of the commonest minerals of the igneous rocks, such as quartz, feldspar, and mica.

None of these ingredients are very soluble in pure water; but if the water contains oxygen and carbonic acid gas in solution, the feldspar will be gradually broken up, and the hard granites or gneisses, which form so large a portion of the igneous rocks, will be gradually disintegrated. From the feldspar will be derived the kaolins or clays, and the water percolating through them will be charged with a small quantity of potash, so necessary for the growth of plants.

Limestones are readily disintegrated by water which contains carbonic acid in solution, and from such rocks are derived the calcareous matter so necessary to plant-growths.

The carbonic acid dissolved in water sometimes acts to considerably change the character of the soil, by effecting new combinations of its mineral constituents.

Some soils possess the very valuable property of absorbing water vapor directly from the atmosphere and condensing it in their pores. This property is exceedingly valuable during times of extended droughts, when a considerable quantity of vapor may be present in the air. Of all soils, those containing the greatest quantity of vegetable mould or humus possess this valuable property in the greatest degree. Clayey soils also possess it to a marked degree.

Soils also possess the power of absorbing gases. Ordinarily, most soils contain of absorbed gases a smaller proportion of oxygen and more carbonic acid gas than the atmosphere.

The ability of soils to absorb the sun's heat varies with their color; as a rule, dark-colored soils absorb the heat more rapidly than light-colored soils.

Darwin has shown that in certain localities the common earthworm greatly aids in the formation and physical character of the soil by extensive burrowing and tunnelling.

In all cases, however, the process by which the soil is formed is a gradual one; and since there is necessarily mingled with such finely broken-up mineral matters a quantity of vegetable humus derived from the decay of successive generations of plants, its formation is rendered still more gradual.

In wooded districts, where a carpet of decaying leaves covers a large part of the ground during most of the year, the water that soaks into such porous soil naturally contains a larger quantity of carbonic acid than would water that had not previously been passed through such matter. This dissolved carbonic acid, from its chemical action on the ingredients of the rock, greatly aids the water in forming new soil.

Speaking of the loss of the vegetable mould, gained by the patient accumulation of different generations of plants through passing centuries, M. de Bouville, a Prefect of the lower Alps, in a report to the Government, quoted by Marsh on page 540 of " The Earth as Modified by Human Action," * writes as follows :

---

* Reprinted, by permission, from " The Earth as Modified by Human Action," by George P. Marsh. New York : Scribner, Armstrong & Co., 654 Broadway, New York, 1874. Pp. 656.

" It is certain that the productive mould of the Alps, swept off by the increasing violence of that curse of the mountains, the torrents, is daily diminishing with fearful rapidity. All our Alps are wholly, or in large proportion, bared of wood. Their soil, scorched by the sun of Provence, cut up by the hoofs of the sheep, which, not finding on the surface the grass they require for their sustenance, gnaw and scratch the ground in search of roots to satisfy their hunger, is periodically washed and carried off by melting snows and summer storms.

" I will not dwell on the effects of the torrents. For sixty years they have been too often depicted to require to be further discussed, but it is important to show that their ravages are daily extending the range of devastation. The bed of the Durance, which now in some places exceeds a mile and a quarter in width, and, at ordinary times, has a current of water less than eleven yards wide, shows something of the extent of the damage. Where, ten years ago, there were still woods and cultivated grounds to be seen, there is now but a vast torrent; there is not one of our mountains which has not at least one torrent, and new ones are daily forming."

The power of a glacier with its fragments of rocks to erode the valleys through which it passes is thus referred to by Le Conte, in his " Elements of Geology," * on page 51.

---

* Reprinted, by permission, from the " Elements of Geology," by Joseph Le Conte, Professor of Geology and Natural History in the University of California. New York: D. Appleton & Co., 551 Broadway, 1878. Pp. 588.

"When we consider the weight of glaciers and their un-yielding nature as compared with water, it is easy to see that their erosive power must be very great. This is increased immensely by fragments of stone of every conceivable size carried along between the glacier and its bed. These partly fall in at the sides and become jammed between the glacier and the confining rocks, partly fall into the crevasses and work their way to the bed, and partly are torn from the rocky bed itself. The effects of glacier erosion differ entirely from those of water : 1. Water, by virtue of its perfect fluidity, wears away the softer spots of the rock and leaves the harder standing in relief; while a glacier, like an unyielding rubber, grinds both hard and soft to one level. This, however, is not so absolutely true of glaciers as might be supposed. Glaciers, for reasons to be discussed hereafter, conform to large and gentle inequalities of their beds, though not to small ones, acting thus like a very stiffly viscous body. Thus, their beds are worn into very re-markable and characteristic smooth and rounded depressions and elevations called *roches moutonées.* Sometimes large and deep hollows are swept out by a glacier at some point where the rock is softer, or where the slope of the bed changes sud-denly from a greater to a less angle. If the glacier should subsequently retire, water accumulates in these excavations and forms lakelets. Such lakelets are common in old glacier beds."

Geikie thus describes the formation of soil in his " Text-Book of Geology," * on page 339 :

* Reprinted, by permission, from "Text-Book of Geology," by Archibald Geikie, LL.D., F.R.S, Director of the Geo-

" On level surfaces of rock the weathered crust may remain with comparatively little rearrangement until plants take root on it, and by their decay supply organic matter to the decomposed layer, which eventually becomes what we term ' vegetable soil.' Animals also furnish a smaller proportion of organic ingredients. Though the character of the soil depends primarily upon the nature of the rock out of which it has been formed, its fertility arises in no small measure from the commingling of decayed animal and vegetable matter with decomposed rock.

" A gradation may be traced from the soil downwards into what is termed the ' subsoil,' and thence into the rock underneath. Between the soil and the subsoil a marked difference in colour is often observable, the former being yellow or brown, when the latter is blue, gray, red, or other colour of the rock beneath. This contrast, evidently due to the oxidation and hydration especially of the iron, extends downwards as far as the subsoil is opened up by the rootlets and fibres to the ready descent of rain-water. The yellowing of the soil may even be occasionally noticed around some stray rootlet which has struck down farther than the rest, below the general limit of the soil."

---

logical Survey of Great Britain and Ireland, etc., etc. London: Macmillan & Co., 1882. Pp. 971.

VI. THE INANIMATE ENEMIES OF THE FOREST.

Like other forms of animate nature, the forest is compelled to make a continual struggle for existence. In order that it may continue to exist, the conditions requisite for its growth must be maintained, and the influences that oppose such growth must be held in check.

The character of the vegetation that covers any region of the earth is dependent not only on the character of its soil, but also on the peculiarities of its climate; such, for example, as the distribution of the temperature throughout the year, the distribution of the moisture, etc.

There is in the vegetable as in the animal world a veritable struggle for existence. Given a particular character of soil and climate in any locality, the plants that will continue to grow in such locality will be those that are best fitted to exist there naturally.

At first all forms may appear; but some particular form may be so much better suited to the natural conditions of the locality, that it will grow and multiply so rapidly as to choke out of existence all other forms. Even such forms, however, will continue to exist only as long as the conditions continue favorable for their existence.

The dependence of plant life on climatic conditions is, perhaps, more marked in the higher forms of the vegetable world. Trees may, in a certain sense, be regarded as of the highest type of plant life. They rigorously depend on conditions of soil and climate for their continued existence. It is true that when a forest is once formed, and its vigorous growth has almost completely shut out the light from the soil in which it is growing, that the numerous forms of vegetable life which lie ready to spring up, should the sun's rays find free access to them, are prevented from growing. Even if they did appear, and the climatic conditions remained as before, the same struggle for existence would again occur, and the same forest would in all probability finally be reproduced. If, however, the character of the soil be altered, or the climatic or other conditions be changed, the forests would either disappear or be replaced by trees of another character.

The forest has many enemies that are ready at

all times to resist its growth, and even to sweep it out of existence.

The soil is continually undergoing a small change in composition as the different growths of plants appear and disappear.

The earth's climate is at present undergoing but very little change. In the geological past such changes were so far-reaching and severe that they were followed by pronounced changes in both the animal and plant life. The comparatively small changes that have occurred within historical time are, perhaps, rather to be regarded as some of the effects produced by the disappearance of certain forms of plant life, than as the causes of such disappearance.

The exact balance of conditions that permit the continued existence of forests is so delicate, that causes, comparatively insignificant in themselves, may finally produce marked effects.

The enemies of the forest may be divided into two classes:

1. Inanimate.

2. Animate.

The principal inanimate enemies of the forest are:

a. Fire.

b. Winds.

c. Floods.

d. Avalanches.

The principal animate enemies of the forest are :

a. Plants.

b. Animals.

c. Man.

Limiting our consideration, for the present, to the inanimate enemies of the forest, we will discuse the manner in which each of these enemies tends to destroy the forest.

Fire.—The destruction of the forest by fire sometimes results finally in a more complete loss than by any other cause. In case of the intelligent removal of the forest by the axe of the lumberman, only the larger trees are cut down, and the smaller ones that are left, getting more heat, light, and nourishment, grow rapidly, and are, in turn, soon ready for removal by the axe, thus to give place to others.

Fire, however, generally removes both great and small.

While a fire may sometimes increase the growth of forests, as in the case of the pitch pine by the destruction of the less hardy forms of plant life, the destruction of the forest by fire, especially on the slopes of mountains, is often so complete that before a new vegetation can appear, the rapid drainage of the slopes is attended by such a loss of the soil as to render such slopes unfit to reproduce the forest trees for a period of time much longer than the life of the average man.

Forest fires are generally kindled during the drier seasons of the year. Under these conditions the rain which subsequently falls is apt to carry off so much of the soil that, even should the trees again appear, the remaining soil would probably be insufficient in quantity to bring them to maturity.

The causes of forest fires are to be found in camp-fires of the lumberman, the burning of brush, the locomotive spark, the lightning bolt, and, perhaps, at times, to the heating power of the sun's rays, concentrated by nodules of gum or resin, acting as burning-glasses.

In the case of newly-settled countries, fires have been sometimes purposely started, for the purpose of readily obtaining an extended pasturage.

The rapidity with which a forest fire spreads depends, of course, on the character of the trees and the force of the wind. It is also, however, dependent largely on the character of the soil. A rocky or sandy soil permits a fire to spread much more rapidly than a damp soil. Some forests of soft and readily ignitible wood have a covering of moss on the soil, which permits the fire, when once started, to spread with awful rapidity.

Extensive tracts in South America, capable of sustaining dense forests, and originally covered by such, are now prevented from so doing by fires that are systematically started every year, for the purpose of obtaining a new growth of grass for pasturage.

The power of the wind in causing the destruction of the forest is, to a great extent, limited to the edges of the forest. In the midst of the forest, the trees stand so close together that they shield one another from the force of the wind. If, however, an opening is made by the axe of the lumberman, by fire, or by any other cause, the wind may cut a wide swath through the forest, and thus destroy many noble trees.

When rivers overflow their banks, thousands of acres of forest trees are often swept away, and in this manner considerable changes may occur in the general character of such districts.

The timber thus thrown into the river channel often forms accumulations called rafts, which, becoming fixed in certain parts of the stream, tend to retard the free drainage of the country, and often result in marked changes in the river channel. Such rafts are still found in the Mackenzie River, and formerly existed in parts of the Mississippi and the Red Rivers.

The effects of the avalanche in sweeping away entire forests from the mountain slopes are well known. Like the influence of the wind, this effect is at first limited to the edges of the forest. If the forests are preserved, the further movement of the avalanche may be checked. In most mountainous countries, forests skirting villages are preserved by rigorous penal laws.

The protection from the destructive effects of avalanches afforded by forests on mountain slopes is shown in the following statement by Élisée Reclus in a work entitled, " The Earth," * on page 171:

"The protecting woods of Switzerland and the Tyrol used to be defended by the national *bann*, and, as it were, 'tabooed.' They were, and still are, called the Bannwoelder. In the valley of the Andermatt, at the northern foot of the St. Goth-

---

* Reprinted, by permission, from "The Earth," by Élisée Reclus. New York: Harper & Brothers, Publishers, Franklin Square. Pp. 573.

ard, the penalty of death was once adjudged on any man found guilty of having made an attempt on the life of one of the trees which shielded the habitations. Added to this, a sort of mystic curse was thought to hang over this impious action, and it was told with horror how drops of blood flowed when the smallest branch was broken off. It was true enough that the destruction of each tree might perhaps be expatiated by the death of a man."

"The village and the great establishment of the baths at Barèges, in the Pyrenees, used to be menaced every year by avalanches rushing down from an elevation of four thousand feet, at an angle of thirty-five degrees. The inhabitants, therefore, were in the habit of leaving vacant spaces between the two quarters of the Barèges, so as to allow a free passage to the descending masses. Lately, however, they have endeavored to do away with the avalanches by means somewhat similar to those employed by the Swiss mountaineers. They have thrown up banks from ten to twelve feet broad on the sides of the ravines, and have furnished these banks with an edging of cast-iron piles. Basket-work, and, here and there, walls of masonry, protect the young growing trees, which are gradually improving under the protection of these defensive works. In the mean time, until the real trees are strong enough to arrest the course of the snow, the artificial trees have well fulfilled the end they were destined for. In 1860, the year the defensive works were finished, the only avalanche which slid into the ravine did not exceed four hundred cubic yards in bulk; while the masses which used to fall down upon the Barèges sometimes attained to more than ninety thousand yards in volume."

Lyell, in his "Principles of Geology," * on page 440, speaks thus of the rafts in the Mississippi :

"One of the most interesting features in the great rivers of this part of America is the frequent accumulation of what are termed 'rafts,' or masses of floating trees, which have been arrested in their progress by snags, islands, shoals, or other obstructions, and made to accumulate, so as to form natural bridges across the stream. One of the largest of these was called the raft of the Atchafalaya, an arm of the Mississippi, which branches off a short distance below its junction with the Red River. The Atchafalaya, being in a direct line with the general direction of the Mississippi, catches a large portion of the timber annually brought down from the north ; and the drift trees collected in about thirty-eight years previous to 1816 formed a continuous raft, no less than ten miles in length, two hundred and twenty yards wide, and eight feet deep. The whole rose and fell with the water, yet was covered with green bushes and trees, and its surface enlivened in the autumn by a variety of beautiful flowers. It went on increasing till about 1835, when some of the trees upon it had grown to the height of about sixty feet. Steps were then taken by the State of Louisiana to clear away the whole raft and open the navigation, which was effected, not without great labor, in the space of four years."

Dana, in his "Manual of Geology," † on page

---

* "Principles of Geology," by Charles Lyell. London : Murray, 1872. Pp. 671.

† Reprinted, by permission, from a "Manual of Geology,"

657, gives the following description of the raft of
the Red River:

"The quantity of wood brought down by some American
rivers is very great. The well-known natural 'raft' obstruct-
ing the Red River had a length, in 1854, of thirteen miles, and
was increasing at the rate of one and a half to two miles a
year, from the annual accessions. The lower end, which was
then fifty-three miles above Shreveport, had been gradually
moving up stream, from the decay of the logs, and formerly
was at Natchitoches, if not still farther down the stream.
Both this stream and the other carry great numbers of the
logs to the delta."

---

by James D. Dana. New York: Ivison, Blakeman, Taylor.
and Co., Publishers. Trübner & Co., London. Pp. 911.

VII. THE ANIMATE ENEMIES OF THE FOREST.

THE animate or living enemies of the forest are :

1. Plants.
2. Animals.
3. Man.

In classifying the enemies of the forest as animate and inanimate, it should be borne in mind that the animate enemies of the forest often call to their aid the powers of inanimate nature. An example of this is seen in the case of the destruction of forests by fire, which are more frequently started by man than in any other way.

The influence of plants on the destruction of forests is limited mainly to the natural struggle which exists between the different forms of plant life for the possession of the soil. There are, however, many forms of parasitic plants, which, growing on the tallest and most vigorous trees, often in the end cause their destruction.

A disease common in parts of Germany, called

the " schullkrankheit," often affects the pine forests.   Trees attacked by this disease soon present
the appearance of having been burnt over, their
boughs and branches rapidly dying or drying up.
The cause of the disease is not exactly known.   It
has, however, been ascribed to the presence of a
fungous growth.

In some parts of Iowa a fungous growth on the
cottonwood trees has resulted in considerable damage to them.   The fungus appears as an orange-
yellow dust on the lower surfaces of the leaves.

The animal enemies of the forest, like the winds,
running water, or the avalanche, produce their
most marked action on the borders or edges of the
forest.

In the deep recesses of the forest the vegetable
kingdom holds almost undisputed sway.   The life-
giving power of the sun's light, and, to a great
extent, that of its heat, are dissipated by the dense
foliage that almost completely shuts out the light
from the dank, gloomy ground.   Animal life, to
a great extent, is crowded out.   Wherever the
sunlight freely enters, animal life appears in
myriad forms, until at length the forest again
chokes it out of existence.

The animal enemies of the forest are too numerous to be more than merely mentioned. The following are among some of the more important:

Domestic animals, which, when allowed to range freely through the woods, often cause much damage by gnawing at the bark of trees, or, in some cases, by the destruction of the foliage.

Among wild animals, the rodents effect the greatest destruction by devouring the bark, and often completely girdling the trees. Among the worst of the rodents may be mentioned rabbits and mice, which gnaw the bark, or gophers, which eat the roots. Beavers, too, destroy forests, not only by the actual cutting down of the trees, but especially by building dams, and thus, by causing the overflow of the intervale, destroying all its growing timber.

Goats and other animals live largely on the bark of trees. In certain parts of the earth, such, for example, as Assyria, Greece, Italy, Spain, and Morocco, the extensive forests which once covered them have been completely destroyed by the ravages of goats.

In general, insects damage trees by feeding on the parts necessary for growth and reproduction. Some insects damage trees by boring the trunks and branches in order to deposit their

D 7

eggs. In all cases the increase in the destruction produced by insect life can be traced to the indiscriminate and foolish slaughter of the insectivorous birds that formerly held such life in check.

Caterpillars often cause considerable destruction to forest trees.

The caterpillar of the pine bombyx often causes great ravages in the pine forests. In Germany these caterpillars are called pine-spinners, from the great number of cocoons with which they cover the pine-trees. Such caterpillars have been known to completely destroy extensive pine forests. The foresters are often compelled to set fire to portions of the forests in order to prevent the too rapid multiplication of this pest.

Another caterpillar, which from its black and white coating is sometimes called the monk or nun caterpillar, is equally destructive, not only to forest trees like the pine, but also the other forest trees, such as the beech, oak, and birch.

Other caterpillars cause great destruction to the forest by eating the tender buds or the young shoots.

Grasshoppers often cause considerable damage to young trees by devouring the leaves, herbs, and tender shoots.

The larvæ of insects do great damage to trees by boring chambers, or tunnellings, either in the heart-wood or in the layers of new wood which lie directly under the bark. The destructive powers of such larvæ are the more marked, since they work silently and in the dark, and their presence can scarcely be detected until they have caused the death of the tree.

A beetle known as the *typographer* (*Bostrychus typographicus*), from the shape of the galleries it burrows out in the trees, causes much damage to the forests, especially to the spruce-firs. Unfortunately, these insects breed very rapidly, and while in the larva state are capable of withstanding the most severe frosts.

Some species of willows are severely injured by the larvæ of a species of saw-fly, which strip the leaves and injure the tree generally.

Perhaps the best remedy for the ravages of insects in general is to be found in the preservation of insectivorous birds.

The most powerful enemy of the forest, however, is civilized man. The products of the forest are clearly man's right by gift of nature. He is lord of the forest as of the rest of the earth, and is, therefore, entitled to the use of the wood thus grown for him. It is, however, by the abuse and not by the use of nature's lavish gifts that man deranges its economy, and thus brings on himself so much punishment. If he would only be careful to select trees of vigorous growth, and in cutting them down would exercise care that the remaining trees might live; if he would carefully preserve the soil, and hold in check the other enemies of the forest; if he would wisely set aside large portions of the mountain slopes, the natural home of the forest, as areas on which trees should be continually preserved, the earth would yield of her abundance all the wood required for his use.

Referring to the insect enemies of the forest, Hough, in a report to the United States Commissioners of Agriculture,* page 263, in citing a writing of Grandjean, Conservateur des Forêts, says:

"The timber-tree particularly suffering from this cause was the Abies excelsa (D.C.), or common European spruce-fir, and the species of insects that did the injury were the Bostrychus typographicus and the B. chalcographicus, of which the first

---

* Reprinted, by permission, from the "Report on Forestry," submitted to Congress by the Commissioner of Agriculture, by Franklin B. Hough. Washington: Government Printing-Office, 1882. Pp. 318.

attacked the trunk and large branches, and the latter, which was seldom absent, found a lodgment in the smaller branches. Their habits were described as follows :

" When the female of the typographic species is ready to deposit her eggs, which occurs about the middle or latter part of spring, sooner or later, according to the temperature, she penetrates the bark, and bores, almost invariably from below upwards, a gallery that is cut along the outer layer of the sapwood, depositing her eggs, as she advances, on the right side and the left. These are so quickly developed that the first larvæ will have themselves made considerable galleries before the parent has finished. Each of these larvæ digs a separate path of its own, more or less inclined to that made by the mother, and at the end of two or two and a half months they are transformed to a perfect insect, which in turn proceeds to lay a new lot of eggs, and, if favored by the heat of August, these are sometimes found more destructive than the first. This second growth is matured towards the end of September or beginning of October, and will be ready to resume operations in the following spring. In the mean time they pass the winter under the mosses and in the crevices of the bark, where they endure the severest frosts of winter, for the perfect insect is as hardy as its larvæ are tender.

" The number of eggs deposited by one insect varies from twenty to one hundred and twenty or one hundred and thirty, and from this bark we may make some very instructive estimates. Suppose that each laying of sixty eggs produces specimens in which the sexes are equal, one female will have produced thirty others, which would each before the end of the year be represented by eighteen hundred of their kind. Half of these, before the end of the second year, have produced eight hundred and ten thousand females, and by the end of the third year seven hundred and twenty-nine millions of the producing sex, and the forest will have fed one billion five hundred and six million six hundred thousand of the progeny of this one parent."

Concerning the destructive effects of the animal kingdom, Geikie, in his " Text-Book of Geology,"* page 456, writes:

"Many animals exercise a ruinously destructive influence on vegetation. Of the various insect plagues of this kind it will be enough to enumerate the locust, phylloxera, and Colorado beetle. The pasture in some parts of the south of Scotland has, in recent years, been damaged by mice, which have increased in numbers owing to the indiscriminate shooting and trapping of owls, hawks, and other predaceous creatures. Grasshoppers cause the destruction of vegetation in some parts of Wyoming and other Western Territories of the United States. The way in which animals destroy each other, often on a great scale, may likewise be included among the geological operations now under description."

Speaking of the influence of certain insects in destroying forests from over extended districts,

---

* Reprinted, by permission, from a "Text-Book of Geology," by Archibald Geikie, LL.D. London: Macmillan & Co., 1882. Pp. 971.

Pouchet, in his work entitled the " The Universe, or the Infinitely Great and the Infinitely Little," * page 218, says:

" If, when the warm breath of spring drives away the rigor of winter and renews life in the fields, we enter one of the great coniferous woods of Germany, we are astonished at the tumult and activity which prevail in lieu of the silence we went there to seek. Everything is in movement.

" Groups of woodmen, foresters, and overseers move about by hundreds, and stretch away like columns of skirmishers; it is a complete army in the field, which opens out wherever there is a large space, and of which the wings are sometimes lost in the windings of the roads, or hidden by the projection of some hillock. This mass of men always moves in order, distributed in troops commanded by experienced leaders. They are all provided with long weapons, which, at a distance, might be taken for lances.

" Or, if the excursion is made by night, another spectacle awaits us. The whole forest seems on fire. In every part are burning great trees, erect and isolated, like huge threatening torches, the flame of which rises to the clouds and casts a baleful glance on all around. A few foresters, standing in silence, contemplate the progress of the conflagration, and watch its ravages. Lastly, at other times, as a final resource, the entire forest is given up a prey to the flames, and whirl-

---

* Reprinted, by permission, from "The Universe, or the Infinitely Great and the Infinitely Little," by F. A. Pouchet, M.D. New York: Charles Scribner & Co., 1870. Pp. 790.

winds of fire, menacing and dreadful, spread on every side ; a woody region, formerly so fertile, is entirely devoured by fire, and only an immense mountain of charcoal remains of all this mass of wealth.

"We ask, against what formidable enemy such an army of men has been launched ? Who are they going to attack with their rods which they brandish on all sides ? What redoubtable aggressors are the others attempting to stay the march of, with the long trenches they are scooping out ? Why these frightful fires in the middle of the night ? Why this general conflagration ?"

VIII. THE DESTRUCTION OF THE FOREST.

THE removal of the forests from any considerable section of country, in the end, is invariably followed by some or all of the following results:

1. An increase in the frequency with which the rivers in that section of country overflow or inundate their banks.

2. An increase in the frequency and severity of droughts, as witnessed by a marked decrease in the amount of water in the river channels, and by an increase in the frequency with which the springs, in such section of country, either show a marked decrease in their flow or dry up altogether.

3. A rapid loss of the soil from such areas, resulting from the more rapid surface drainage of their surfaces.

4. A marked disturbance in the lower courses of the rivers, rising in or flowing through such section of country, produced by the filling up of their channels by sand-bars or mud-flats.

f

5. A decrease in the healthfulness of the district that borders on the lower courses of such rivers,—that is, in those portions which lie in the lowlands near the rivers' mouths.

6. An increase in the number and severity of hail-storms, both over the areas themselves or in the countries bordering thereon.

When the forests are removed from any section of country, that part of the rainfall which formerly entered the ground, either by gradually sinking into the porous soil, or by running along the branches and trunks of the trees, and so entering and penetrating the more deeply-seated strata, now drains rapidly off the surface. Instead of reaching the river channel quietly and slowly through discharge from the reservoirs of springs, it now rapidly drains directly off the surface into the river channel.

Instead of draining into the river channel continuously for a period of, say, three weeks, the rain-water now drains into the channel in often a period of as many hours. The channel rapidly fills, the river overflows its banks, and the floods so caused carry loss to the lowlands along the river banks, and, not infrequently, death to the inhabitants.

Not only are the riches of the rainfall thus squandered, to the loss of the inhabitants of the river valleys, from the excess of water immediately after a rainfall, but a still greater and more far reaching loss occurs from the failure of the rainfall to fill the reservoirs of the springs, the continuous discharge of which are necessary to maintain the proper flow of water in the river.

The springs, having their reservoirs but partly filled, are apt to fail shortly after the rainfall ceases, so that even limited droughts may cause them to dry up completely.

The damage, however, does not stop here. The soil in which the forest grew, being no longer held together either by the roots of the trees or underbrush of the forest, or protected by a vegetable covering, is rapidly carried away by the water. The soil thus lost, resulted from the gradual disintegration of hard rocks, and contains as essential elements substances derived from the continued growth of former generations of plants, and probably required centuries for its production. Its removal in a few years is, therefore, a serious matter.

The soil, the wealth of the highlands, is now thrown into the river channel, and though some of it fertilizes the lowlands, over which it is spread during inundations, yet much collects in sand-bars and mud-flats on the lower courses of the river.

These flats work injury because:

1. They hinder navigation, and thus interfere with the commerce between different parts of the country.

2. They become sources of contamination to the air of the lowlands, by breeding miasmatic and other diseases.

Besides the disturbances thus caused to the drainage of the region from which the forest has been removed, considerable changes are brought about in the rate at which the now bare soil receives the heat from the sun, and the rapidity with which it throws it off into the air.

Areas covered with forests both receive and part with their heat slowly, and are, therefore, not very apt to become very hot in summer, or very cold in winter.

Bare areas, or areas stripped of their vegetable covering, both receive and part with their heat rapidly, and are, therefore, apt to become very hot in summer and very cold in winter.

The presence of the forest, therefore, tends to prevent marked changes in the temperature of the air, while the removal of the forest tends to permit sudden changes in such temperature.

These effects will be considered under the general head of climate.

The axe of the pioneer, so often regarded as the emblem of civilization, is more correctly to be regarded as an emblem of an entirely different character.

The problem of the preservation and protection of the forest is one of extreme difficulty, for the following reasons :

The dense populations which now exist in most of the temperate regions of the earth could not continue to exist in the forest regions which once grew on large parts of their areas.

The regions best fitted for the growth of men are also best fitted for the growth of trees. Since civilized man cannot continue as a dweller in the forest, as the density of population increases, the forest must be cut down.

In removing the forest to make way for man, certain areas should be set aside in all sections for the purpose of perpetually maintaining trees thereon. The nature of such areas will, of course, depend on a variety of circumstances. In general, however, it can be shown that, on the slopes of mountain ranges, which form the natural places where rivers rise, forests should be especially maintained.

Laws should, therefore, be enacted providing for the replanting of trees on mountain slopes, either when they have been removed by the axe of the woodman, or by fire, or by any of the other enemies of the forest.

The influence of the destruction of the forest on the rapidity of drainage, and the consequent liability to the destructive floods, is thus referred to by the author in his "Elements of Physical Geography," * page 64, as follows:

"*Influence of the Destruction of the Forests on Inundations.*— When the forests are removed from a large portion of a river-basin, the rains are no longer absorbed quietly by the ground, but drain rapidly off its surface into the river channels, and thus in a short time the entire precipitation is poured into the main channel, causing an overflow. It is from this cause that the disastrous effects of otherwise harmless storms are produced. The inundations are most intensified by this cause in the early spring, when the ice and snow begin to melt. The destructive effects of the floods are increased by the masses of

---

* Reprinted, by permission, from "The Elements of Physical Geography," by Edwin J. Houston, A.M. Philadelphia: Eldredge & Brother, No. 17 North Seventh Street, 1891. Pp. 172.

floating ice, which, becoming gorged in shallow places in the stream, back up the waters above. The increased frequency of inundations in the United States is, to a great extent, to be attributed to the rapid destruction of the forests."

Sir Charles Lyell, in his "Principles of Geology," * speaking of the effects produced by the removal of the forest, says, on page 457:

"When St. Helena was discovered, about the year 1506, it was entirely covered with forests, the trees drooping over the tremendous precipices that overhang the sea. Now, says Dr. Hooker, all is changed; fully five-sixths of the island is entirely barren, and by far the greater part of the vegetation that exists, whether herbs, shrubs, or trees, consists of introduced European, American, African, and Australian plants, which propagated themselves with such rapidity that the native plants could not compete with them. These exotic species, together with the goats, which, being carried to the island, destroyed the forests by devouring all the young plants, are supposed to have utterly annihilated about one hundred peculiar and indigenous species, all record of which is lost to science, except those of which specimens were collected by the late Dr. Burchell, and are now in the herbarium of Kew."

The protective action or plants generally as preventing erosion by water or wind is clearly pointed

* "Principles of Geology," by Sir Charles Lyell, M.A. London: John Murray, 1872. Pp. 652.

out by Geikie, in his "Text-Book of Geology,"*
on page 456.

"The protective influence of vegetation is well known.

"1. The formation of a stratum of turf protects soil and rocks
from being rapidly removed by rain or wind. Hence, the
surface of a district so protected is denuded with extreme
slowness except along the lines of its water-courses.

"2. Many plants, even without forming a layer of turf, serve
by their roots or branches to protect the loose sand or soil on
which they grow from being removed by wind. The common
sand-carex and other arenaceous plants bind littoral sand-
dunes and give them a permanence which would at once be
destroyed were the sand laid bare again to storms. In North
America the sandy tracts of the Western Territories are in
many places protected by the sage-brush and grease-wood.
The growth of shrubs and brushwood along the course of a
stream not only keeps the alluvial banks from being so easily
undermined and removed as would otherwise be the case, but
serves to arrest the sediment in floods, filtering the water and
thereby adding to the height of the flood-plain. On some parts
of the west coast of France extensive ranges of sand-hills have
been gradually planted with pine woods, which, while preventing the destructive inland march of the sand, also yield a large
revenue in timber, and have so influenced the climate as to
make these districts a resort for pulmonary invalids. In tropi-

cal countries the mangrove grows along the sea-margin, and not only protects the land, but adds to its breadth, by forming and increasing a maritime alluvial belt."

The following, from the "Journal of the Society of Arts," * shows the enormous demands made on the forest by railroads for sleepers :

"The Belgian 'Bulletin du Musée Commercial' gives the following information respecting the number of sleepers used on various railways. In France alone the six larger railway companies require a daily supply of more than ten thousand sleepers, making an annual consumption of over three million six hundred and fifty thousand. As a tree of ordinary dimensions cannot furnish more than ten logs, it follows that more than a thousand fine trees are cut down every day solely for the purpose of supplying the necessary sleepers for the French railways. In the United States the amount required is still greater. Over fifteen million sleepers are annually used in this country, thus necessitating the annual destruction of eighty thousand hectares, or one hundred and ninety-seven thousand six hundred acres of forests. 'The Bulletin du Musée Commercial' estimates at more than forty millions the number of logs required for the railways of the world, and is of opinion that the estimate is rather below than above the mark."

---

* "Journal of the Society of Arts," vol. xxvii. London : George Bell & Sons, 6 York Street, Covent Garden, 1889. Pp. 924.

IX. THE EARTH'S OCEAN OF VAPOR.

FROM every water surface on the earth there is almost constantly rising and passing into the air an invisible form of water called vapor.

Vapor is formed wherever water is sufficiently heated under such circumstances that its particles have freedom to expand, and thus occupy a greater space.

The waters of the earth are caused to pass into the atmosphere as vapor mainly by the heat of the sun.

The vapor that passes into the air from the ocean and other water surfaces spreads or diffuses through the air, and is carried by the winds to different parts of the earth's surface. The air directly over a water surface is, however, generally moister than that over a land surface.

When, by any cause, water vapor loses the heat which caused it to become a vapor, it again becomes visible as dew, fog, cloud, or mist, or falls as rain, hail, or snow.

The rapidity with which water surfaces throw

off vapor into the air varies with the following circumstances :

1. With the amount of surface exposed.

Evaporation takes place only at the surface; consequently, the greater the surface, the greater the rapidity of evaporation. When wet clothes are hung out to dry, they are so opened or spread out that the air can act on them from all sides. A pound of water placed in an open shallow dish, and exposed to the air, will evaporate much more rapidly than the same quantity would if placed in an open, narrow-necked bottle.

For the same reason, an equal quantity of water will evaporate still more rapidly when sprinkled on the surface of a sheet hung out in the air to dry.

2. On the temperature of the air.

The capacity of a given volume of air for water in a state of vapor rapidly increases with its temperature. A cubic foot of dry air at the temperature of melting ice, or 32 degrees Fahrenheit, when saturated, holds a little more than half a grain of vapor. It then being saturated can hold no more water in an invisible state. Increase its temperature, however, to 212 degrees Fahrenheit, and it can hold twenty grains, or about forty times as much as it formerly held.

Consequently, any increase in the temperature of air permits it to hold a greater quantity of vapor. Conversely, any decrease in the temperature of air causes its ability to hold moisture as vapor to decrease.

If, therefore, the temperature of the air be sufficiently decreased, a part of the vapor it contains will appear in some visible form.

3. On the quantity of vapor already in the air.

When a given bulk of air has as much vapor in it as it can hold, all evaporation ceases. Consequently, the drier the air over a water surface, the greater is the rapidity of evaporation.

4. On the velocity of the wind.

The wind brings fresh and drier air to the water surfaces, and at the same time removes the air into which such surfaces were discharging their vapor. An increase in the velocity of the wind, therefore, increases the rapidity of evaporation.

5. On the pressure of the air.

The greater pressure the air exerts on a water surface, the slower the rapidity of evaporation. A low barometer permits a water surface to throw off its vapor with much greater rapidity than a high barometer.

When clouds reach the higher regions of the atmosphere they disappear, because the particles of water of which they are composed pass rapidly into an invisible vapor, on account of the great relief of pressure in the higher regions.

The earth's ocean of vapor is of the greatest importance to its present race of animals and plants. If any considerable change in either the quantity or distribution of vapor should be maintained for any considerable time, the present race of animals and plants would disappear.

Some of the more important ways in which the ocean of vapor affects the economy of the earth are as follows:

1. By the action of the winds, the water vapor is carried from the warm regions of the earth to the colder regions, where, falling as rain or snow, it gives out its heat and raises the temperature of the air over such regions.

An interchange is thus effected between the too great heat of the equatorial regions and the too feeble heat of the poles, and a more equable, uniform temperature is insured than would otherwise exist.

The earth's ocean of vapor therefore acts to moderate the excessive temperatures that would otherwise exist both in the equatorial and in the polar regions.

2. By acting as a screen interposed between the earth and the sun, and thus preventing the earth's surface from becoming too rapidly heated when exposed to the sun's rays, or too rapidly cooled when deprived of such rays.

Water enters so largely into the composition of both animals and plants, that its absence from any section of country invariably causes such section to become a desert.

Within certain limits, the wealth of any section of country can be accurately estimated by the number of inches of rain that fall in a given time on its surface. This liquid wealth may be regarded as a species of bank account of such section of country, by which its solvency or bankruptcy may be determined.

The ocean of vapor which forms the source from which the rains are derived is, therefore, of great importance to the operations of nature.

Even a hurried glance at the map of the world will show that the earth's greatest expanse of water surface occurs near the equatorial regions. Here, also, the sun's heat is greatest. The air over the equatorial regions would become too enormously heated to sustain the present life of the earth, were the water surfaces replaced by land.

Not only does a water surface heat more rapidly than a land surface, but the vapor which arises from it, locks up much of the heat in a form that is sometimes popularly called latent heat.

To change a pound of ice, at thirty-two degrees Fahrenheit, into a pound of water, at thirty-two degrees Fahrenheit, requires one hundred and forty-two heat units, or one hundred and forty-two times as much heat as is required to raise the temperature of a pound of water one degree Fahrenheit. To convert one pound of water at sixty degrees Fahrenheit into vapor requires nearly one thousand heat units, an amount of heat that would be able to raise more than six pounds of ice-cold water to the temperature of its boiling-point. When the vapor is condensed and falls as rain or snow, this heat reappears and raises the temperature of the air. When, therefore, the excessive heat of the sun in the equatorial regions falls on the extended water surfaces, much of the heat is absorbed by the vapor, and the air is prevented from growing too hot. This vapor is carried by the winds to the polar regions, where it gives up its heat to the air, and falls as rain or snow.

The vapor of water exerts another and still more powerful influence on the climate of the earth. Water vapor possesses in a marked degree the power of absorbing heat rays of the sun. About twenty-eight per cent. of the heat of the vertical rays is absorbed before such rays reach the surface, provided there is a sufficient quantity of vapor in the air. When the heated earth throws off or radiates its heat into the atmosphere, the same water vapor absorbs a greater part of such rays, and rapid cooling by radiation is thus prevented. The presence of the water vapor, therefore, prevents either the rapid heating of the earth's surface by the direct action of the sun's rays, or the rapid cooling of such surface by radiation.

If the earth's surface were deprived of this screen of vapor, the air would become so rapidly heated on the rising of the sun, and so rapidly cooled on its setting, that the earth would be unable to sustain its present plant and animal life.

Tyndall, speaking of the influence that the earth's water vapor exerts on the climate of England, says, " The removal for a single summer night of the clouds of vapor which cover England would be attended by the destruction of every plant which a freezing temperature could kill."

The amount of vapor in the air of any country, though dependent on the direction from which the winds come, is also markedly influenced by the nature of its surface.

The presence of forests over any section of country has the effect of decreasing the rapidity with which the wet surface parts with or loses its water by evaporation. This decrease in the rapidity of evaporation is caused :

(*a.*) Because the air over the forest is generally moister than that over the open fields, and evaporation takes place less rapidly in moist air.

(*b.*) Because the ground in the forest is shielded from the direct rays of the sun.

(*c.*) Because the wet ground is protected from the direct action of the wind.

The presence of the forest, therefore, tends to keep the air moist for a longer time, and to thus prevent the occurrence of marked contrasts in the humidity of the air.

Some experiments made in France show that the rapidity of evaporation is sixty-three per cent. less in the forest than in the open fields.

The influence of the vapor screen that is placed between any surface and the sun on the climate of the surface is thus referred to by Tyndall in his " Heat as a Mode of Motion," * on page 417 :

" A few years ago a work possessing great charms of style and ingenuity of reasoning was written to prove that the more distant planets of our system were uninhabitable. Applying the law of inverse squares to their distances from the sun, the diminution of temperature was found to be so great as to preclude the possibility of human life in the more remote members of the solar system. But in those calculations the influence of an atmospheric envelope was overlooked, and this omission vitiated the entire argument. An atmosphere may act the part of a barb to the solar rays, permitting them to reach the earth, but preventing their escape. A layer of air two inches in thickness, saturated with the vapor of sulphuric ether, would offer very little resistance to the passage of the solar rays, but I find that it would cut off fully thirty-five per cent. of the planetary radiation. It would require no inordinate thickening of the layer of vapor to double this absorption; and it is perfectly evident that, with a protecting envelope of this kind, permitting the heat to enter, but preventing its escape, a comfortable temperature might be . obtained on the surface of the most distant planet."

---

* Reprinted, by permission, from " Heat as a Mode of Motion," by John Tyndall, LL.D., F.R.S.   New York: Appleton and Company, 1883.   Pp. 591.

Alexander von Humboldt, in his " Cosmos," *
thus refers to the vapor of the atmosphere, on
page 330, of vol. i. :

"As the quantity of moisture in the atmosphere increases
with the temperature, this element, so important to the whole
organic creation, varies with the hour of the day, the season
of the year, and the degree of latitude and of elevation. Our
knowledge of the hygrometric relations of the atmosphere has
been materially augmented of late years by the method now
so generally and extensively employed of determining the
relative quantity of vapor, or the conditions of moisture of
the atmosphere, by means of the difference of the *dew point*
and of the temperature of the air, according to the ideas of
Daniell and of Dalton, and by the use of the wet-bulb thermometer. Temperature, atmospheric pressure, and the direction of the wind have all a most intimate relation to the
atmospheric moisture so essential to organic life. The influence, however, of humidity on organic life is less a consequence of the quantity of vapor held in solution under
different zones than the nature and frequency of the aqueous
precipitations which refresh the ground in the form of dew,
mist, rain, or snow."

---

* "Cosmos," vol. i., by Alexander von Humboldt.   London : Longman, Brown, Green & Longmans, 1849.   Pp. 487.

X. RAIN.

THE vapor which rises from the surface of the ocean, and, indeed, from all water surfaces, mixes or diffuses through the air, and is carried by the winds to different parts of the earth. The air over parts of the earth at considerable distances from any large body of water may therefore contain much vapor.

The quantity of water the air can hold, in an invisible state as vapor, increases rapidly with an increase in the temperature. Consequently, when air containing vapor is considerably chilled, it can no longer hold as much as it formerly did, and a part appears as rain, or as some other form of precipitation, such as dew, snow, hail, fog, cloud, etc.

The amount of water that falls, or is precipitated from the air, depends not only on the quantity of air that is chilled, and on the extent of this chilling, but also on the quantity of moisture the air contained before it was chilled.

The lowering of temperature necessary to produce rain may be caused in the following ways:

1. The moist air may blow along the earth's surface towards colder regions.

2. The moist air may rise directly from the earth's surface into the higher and colder regions of the air.

As a rule, the moist air which blows along the earth's surface towards the poles becomes chilled and deposits its moisture as rain or snow. On the contrary, the moist air which blows along the earth's surface towards the equator becomes, for the greater part, warmer and, thus, becoming drier takes rather than gives moisture, and produces drought.

Therefore, as a rule, only the surface winds which blow towards the colder regions of the earth can be expected to bring rain.

In the tropical regions, however, any wind, whether from the equator or from the poles, which has crossed the ocean or any other large body of water, and has thereby become saturated with moisture, will deposit some of its moisture as rain when it strikes the cooler coasts of a continent or island. Even in such cases, however, the equatorial winds are more apt to cause heavy rainfalls than those from the poles.

A warm, moist air, when sufficiently chilled, will

cause a heavier rainfall than a cold, moist air, because the warm air has a greater capacity for holding vapor.

In general, the air of the equatorial zones of the earth is both warmer and moister than that of the temperate zones, and the air of the temperate zones is both warmer and moister than that of the polar zones.

Consequently the rainfall is heaviest in the equatorial zones, and is greater in the temperate zones than in the polar zones.

The air near the coast of a continent or island is moister than that over the interior. Consequently the rainfall is heavier on the coasts than in the interior.

When the earth's surface is intensely heated, the air over it becomes so hot that it rises far above the surface. If sufficiently moist, the chilling so caused produces a heavy rainfall. Much of the rain in the tropical regions is caused in this manner.

Mountains form excellent means for cooling the air and causing its invisible water or vapor to fall as rain. They act no matter from what direction the wind may be blowing.

When the wind blows against the sides or slopes of a mountain, it is forced by the pressure of the wind behind it to slowly creep up the slopes of the mountain, and becomes chilled in the colder regions which lie near the summit. If this lowering of temperature be sufficiently great, the moisture will be precipitated from the air, no matter from what direction the wind may come.

Mountains may, therefore, cause rain to fall from any wind that is forced to blow over them, provided they are sufficiently high to cause the necessary amount of cooling. When a mountain reaches sufficiently far upward into the air to cause the temperature to fall below the freezing-point of water, the condensed moisture falls as snow.

The reason so many rivers rise in mountains is to be found in the fact that the mountains act to chill the winds, and so rob the air of its moisture, no matter in from what direction the wind, which is forced to ascend their slopes, may happen to blow.

Nearly all the rivers of the world rise in mountainous districts. As a rule, the largest rivers of the world rise in the highest mountains. This is because the higher the mountain the colder its slopes, the cold mountain slopes acting, as explained, to deprive the air of its moisture.

The rain that falls on a mountain's slopes, like

that which falls on any other part of the earth's surface, either runs rapidly off the surface or sinks slowly into the ground.

The part that runs directly off the ground will be greater than the part which sinks into the ground, when the surface is bare and devoid of vegetation. On the contrary, the part which sinks into the ground will be greater than the part which runs directly off the surface, when the surface is covered by forests. But the proportion of the rainfall which sinks into the ground, as compared with that which runs directly off the surface, is greater where the sides of mountains are covered with forests than in any other case.

Since the rivers which rise in the mountains are more regularly fed by the springs when the greater part of the rainfall sinks quietly into the ground, and since this occurs on mountains that are covered with trees, the importance of keeping the sides of the mountains well wooded is evident.

When the sides of mountains are covered with forests, the rivers that rise on their slopes are not only less apt to overflow their banks during heavy rainfalls, but are also less apt to dry up and become shallow during droughts, than if such forests were removed.

The forests should, therefore, be preserved on the mountain-sides, in order to protect the lowlands either from inundations or floods, or from the effects of too small a quantity of water in the rivers which flow through them during droughts.

The action of mountains in cooling the air and causing the condensation of the moisture of the air, is thus referred to by Tyndall in his " Heat as a Mode of Motion," * page 384 :

"Mountains act as condensers, partly by the coldness of their own masses, which they owe to their elevation. Above them spreads no vapor screens of sufficient density to intercept their heat, which, consequently, passes unrequited into space. When the sun is withdrawn, this loss is shown by the quick descent of the thermometer. The difference between a thermometer which, properly protected, gives the true temperature of the night-air, and one which is permitted to radiate freely towards space, must be greater at high elevations than at low ones. This conclusion is confirmed by observation. On the Grand Plateau of Mont Blanc, for example, MM. Martins and Bravais found the difference between two such thermometers to be twenty-four degrees Fahrenheit, when a difference of only ten degrees was observed at Chamouni."

---

* Reprinted, by permission, from "Heat as a Mode of Motion," by John Tyndall, LL.D., F.R.S. New York: D. Appleton & Co., 1883. Pp. 591.

Huxley, in his "Physiography," * speaks as follows, concerning the formation of rain, on page 47:

"In examining the distribution of rain, it will be found to be regulated partly by the physical features of the country, and partly by the character of the prevailing winds. In the neighborhood of mountains, the rainfall is increased, since, as has already been pointed out, a mass of moist air, when forced up the side of a mountain, is chilled in the ascent, and its moisture consequently discharged. Among our western counties, in the neighborhood of hills, the rainfall rises to eighty, or even to a hundred, inches, and upwards; while away from hills, though still in the west, it is only from thirty to forty-five inches. A table-land, or high plain surrounded by mountains, will generally receive but little rain, since the winds which reach it have been more or less drained of moisture in sweeping over the surrounding hills. For a like reason, but little rain is likely to fall on the lee side of a high hill, and many mountains, consequently, have a wet and a dry side; the wet side being, of course, that towards which the predominant winds blow. As regards the influence of winds on rain, it is evident that, when air has blown over a large expanse of warm water, it must have become laden with moisture, which will be readily precipitated on exposure to refrigerating influences. Hence, as in Britain, so in the greater part of Europe, the southerly and westerly winds bring rain; and most rain

---

* Reprinted, by permission, from "Physiography," by T. H. Huxley, F.R.S. London: Macmillan & Co., 1883. Pp. 384.

falls in the exposed westerly parts, such as the coast of Portugal, Spain, France, Britain, and Norway. There are certain conditions, however, under which rain is brought to our islands by easterly rather than westerly winds."

Maury, in his "Physical Geography of the Sea," * on page 120, says :

"We shall now be enabled to determine, if the views which I have been endeavoring to present be correct, what parts of the earth are subject to the greatest fall of rain. They should be on the slopes of those mountains which the trade-winds or monsoons first strike after having blown across the greatest tract of ocean. The more abrupt the elevation, and the shorter the distance between the mountain-top and the ocean, the greater the amount of precipitation. If, therefore, we commence at the parallel of about thirty degrees north in the Pacific, where the northeast trade-winds first strike that ocean, and trace them through their circuits till they meet high land, we ought to find such a place of heavy rains. Commencing at this parallel of thirty degrees, therefore, in the North Pacific, and tracing thence the course of the northeast trade-winds, we shall find that they blow thence, and reach the region of equatorial calms near the Caroline Islands. Here they rise up ; but, instead of pursuing the same course in the upper

---

* Reprinted, by permission, from "The Physical Geography of the Sea, and its Meteorology," by M. F. Maury, LL.D., U.S.N. New York : Harper & Brothers, Publishers, Franklin Square.

stratum of winds through the southern hemisphere, they, in consequence of the rotation of the earth, are made to take a southeast course. They keep in this upper stratum until they reach the calms of Capricorn, between the parallels of thirty degrees and forty degrees, after which they become the prevailing northwest winds of the southern hemisphere, which correspond to the southwest of the northern. Continuing on to the southeast, they are now the surface winds; they are going from warmer to cooler latitudes; they become as the wet sponge, and are abruptly intercepted by the Andes of Patagonia, whose cold summit compresses them, and with its low dew-point squeezes the water out of them. Captain King found the astonishing fall of water here of nearly thirteen feet (one hundred and fifty-one inches) in forty-one days; and Mr. Darwin reports that the sea-water along this part of the South American coast is sometimes quite fresh, from the vast quantity of rain that falls. A similar rainfall occurs on the sides of Cherraponjie, a mountain in India. Colonel Sykes reports a fall here during the southwest monsoons of six hundred and five and one-quarter inches. This is at the rate of eighty-six feet during the year; but King's Patagonia rainfall is at the rate of one hundred and fourteen feet during the year. Cherraponjie is not so near the coast as the Patagonia range, and the monsoons lose moisture before they reach it."

XI. DRAINAGE.

THE rain that falls on the earth either runs directly off the surface or sinks into the ground.

The part that runs directly off the surface collects in small streams that discharge through a river, either into a lake or into the ocean.

The part which sinks into the ground collects in pockets or places below the surface, called reservoirs. As a rule, the water escapes from these underground reservoirs by coming out at the surface at some lower level, as a spring. During most of the time the flow of water in a river is kept up by the springs pouring their waters into the many streams that empty into the river channel.

The water, therefore, that falls from the sky as rain, flows directly from the earth's surface into a river, or first collects in a reservoir, from which it afterwards flows into a river.

The running of the water from the level where the rain fell to a lower level is called drainage.

There are two kinds of drainage:

1. Surface drainage, or where the rain-water runs directly off the surface.

2. Underground drainage, or where the rain-water first sinks into the ground and then discharges as springs into some stream that empties into a river.

Surface drainage, for the greater part, takes place rapidly, and occurs mainly during the time rain is falling. It practically stops a few hours after the rain ceases.

Underground drainage takes place slowly, and may continue for many weeks after the rain ceases.

All the water in a river comes from the rain that falls on the earth's surface. The rivers continue to flow because the springs are continually emptying their waters into the rivers, and, before they run dry, more rain falls and keeps up the supply in their reservoirs.

Some rivers are larger than others. This is because :

1. More rain falls on those parts of the earth through which they flow.

2. The land which slopes towards such rivers covers a greater part of the earth's surface.

The water runs off the earth from a higher to a lower level, because water runs down hill. The direction in which water will drain from the land will depend on the direction of the slope of the land. If a large area of land so slopes that all the water draining from it collects in streams flowing into the ocean through a common river mouth, and the rainfall on such area is large, the river itself will be large.

The smaller streams and rivers which collect in a single and larger river, and discharge their waters through a common mouth, are called, collectively, a river system.

The area of land that drains into a river is called a river basin.

The size of a river, therefore, depends upon the amount of the rainfall on its basin, and on the size of its basin.

When the quantity of water discharged into a river is greater than its channel can hold, a flood occurs, or the river is said to inundate its banks.

A heavy rain-fall does not necessarily produce an inundation. If the character of the river basin is such that a comparatively small part of the rain-fall runs directly off the surface, and a large part sinks into the ground and collects in the reservoirs of springs, and slowly passes through such springs into the rivers, sufficient time may be given for the river to safely discharge the waters of even a very heavy rainfall.

If, however, the character of the surface is such that the larger part of the rainfall runs directly off the slopes into the river channel, then an inundation must necessarily attend every heavy rainfall.

If the greater part of the rainfall runs directly off the surface into the river channel, and a comparatively small part goes to feed the reservoirs of the springs, and if a long time elapses before the next rainfall, the springs will dry up, and the water in the river will get very low.

Any disturbance in the natural drainage of a country may cause a damage of two different kinds:

1. The damage due to the overflowing of the rivers, or that directly due to too much water.

2. The damage due to the drying up, or the getting too low, of the rivers in the intervals between the storms, or that due to too little water.

The proportion of the rainfall that sinks quietly into the earth, as compared with that which flows directly off its surface, depends on the character of the surface. As a rule, a surface devoid of vegetable covering—that is, a surface on which no vegetation is growing—will permit a larger proportion of the rainfall to drain directly into the river channels than will a surface covered by vegetation. This is especially the case during the colder parts of the year, when the ground is frozen.

When rain falls on a surface covered by vegetation, the water, by slowly trickling down the stalks or stems of the leaves and the branches and trunks of the trees, finds a ready entrance into the ground by following their surfaces and discharging into the porous ground lying around their roots.

A forest permits this action of the water in sinking into the ground to take place quite readily.

A forest, therefore, tends to decrease the amount of rainfall that drains directly from the earth's surface.

A forest also tends to prevent the occurrence of too little water in a river, because it insures the filling of the reservoirs of springs, which discharge their waters into the rivers during the intervals between the rainfalls.

Unless, therefore, forests are preserved, the proper drainage of the earth will be disturbed, and the rivers will have too much water in their channels during the time of rains, and too little water in the intervals between rains.

The rapid drainage of the surface when no

longer protected by the forests is thus described
by Sir Charles Lyell, in his "Principles of Geology, or the Modern Changes of the Earth and
its Inhabitants," * on page 338, vol. i. :

"When travelling in Georgia and Alabama in 1846, I saw
in both these States the commencement of hundreds of valleys in places where the native forests had recently been removed. One of these newly-formed gulleys or ravines is
represented in the annexed wood-cut, from a drawing which
I made on the spot. It occurs three miles and a half due west
of Milledgeville, the capital of Georgia, and is situated on the
farm of Pomona, on the direct road to Macon.

"In 1826, before the land was cleared, it had no existence ;
when the trees of the forest were cut down, cracks three feet
deep were caused by the sun's heat in the clay ; and during
the rains, a sudden rush of water through the principal crack
deepened it at its lower extremity, from whence the excavating
power worked backward, till, in the course of twenty years, a
chasm measuring no less than fifty-five feet in depth, three
hundred yards in length, and varying in width from twenty to
one hundred and eighty feet, was the result. The high road
had been several times turned to avoid this cavity, the enlarge-
ment of which is still proceeding, and the old line of road may
be seen to have held its course directly over what is now the
widest part of the ravine. In the perpendicular walls of this

---

* "Principles of Geology," by Sir Charles Lyell, F.R.S.,
M.A. London : John Murray, Albemarle Street, 1872. Pp.
650.

great chasm appear beds of clay and sand, red, white, yellow, and green, produced by the decomposition *in situ* of hornblendic gneiss with layers and veins of quartz, which remains entire to prove that the whole mass was once crystalline."

Marsh, in his book on "The Earth as Modified by Human Action," * in referring to the effects produced on the drainage of the land by the destruction of the forest, on page 254, gives the following quotation from a paper read by Blanqui, read before the Academy of Moral and Political Science in 1843, concerning the Alps of Provence:

"The Alps of Provence present a terrible aspect. In the more equable climate of Northern France, one can form no conception of those parched mountain gorges where not even a bush can be found to shelter a bird, where, at most, the wanderer sees in summer here and there a withered lavender, where all the springs are dried up, and where a dead silence, hardly broken by even the hum of an insect, prevails. But if a storm bursts forth, masses of water suddenly shoot from the mountain heights into the shattered gulfs, waste without irrigating, deluge without refreshing the soil they overflow in their swift descent, and leave it even more seared than it was from want of moisture. Man at last retires from the fearful

* Reprinted, by permission, from "The Earth as Modified by Human Action," by George P. Marsh. New York: Scribner, Armstrong & Co., No. 654 Broadway, 1874. Pp. 656.

desert, and I have, the present season, found not a living soul in districts where I remember to have enjoyed hospitality thirty years ago."

The influence of a vegetable covering on the drainage of the surface is thus referred to by Élisée Reclus, in his work on "The Earth : A Descriptive History of the Phenomena of the Life of the Globe," * on page 223 :

"The action of vegetation is not confined merely to imbibing the water falling from the clouds ; it often, also, assists the superabundant moisture in penetrating the interior of the ground. Trees, after they have received the water upon their foliage, let it trickle down drop by drop on the gradually softened earth, and thus facilitate the gentle permeation of the moisture into the substratum ; another part of the rain-water, running down the trunk and along the roots, at once finds its way to the lower strata. On mountain slopes, the mosses and the freshly-growing carpet of Alpine plants swell like sponges when they are watered with rain or melted snow, and retain the moisture in the interstices of their leaves and stalks until the vegetable mass is thoroughly saturated and the liquid surplus flows away. Peat-mosses especially absorb a very considerable quantity of water, and form great feeding-reservoirs for the springs which gush out at a lower level. The immense

---

* Reprinted, by permission, from "The Earth," by Élisée Reclus. New York: Harper & Brothers, Franklin Square. Pp. 573.

fields of peat which cover hundreds and thousands of acres on the mountain slopes of Ireland and Scotland may, notwithstanding their elevation and inclined position, be considered as actual lacustrine basins containing millions of tons of water dispersed among their innumerable leaflets. The superabundant water of these tracks of peat-mosses issues forth in springs on the plains below."

The protective action of a vegetable covering is thus alluded to by Prestwich in his " Geology, Chemical, Physical, and Stratigraphical," * page 136 :

" This surface soil, with its usual covering of herbage, serves to protect the land from further degradation, and checks the denuding action which would otherwise scour the surface after every shower of rain. Instances have been adduced to show how persistent are the features of such a surface. The positions of the many dolmens and other so-called ' Druidical' stones, so common on the downs of this country and in many parts of France, shows that the level of the vegetable soil has undergone little or no change since they were first erected. The camp of Attila, situated in the great chalk plains of Champagne, furnishes a well-known date, namely, A.D. 451. Notwithstanding its more than fourteen hundred years, the

---

* Reprinted, by permission, from "Geology, Chemical, Physical, and Stratigraphical," by Joseph Prestwich, M.A., F.R.S. Vol. i. Oxford, Clarendon Press, 1886. Pp. 477.

surface of this great earthwork, which is merely covered with a thin growth of grass, remains almost as perfect and as sharp as when first made and grassed over. Nothing of importance has been removed from the surface by mechanical means, whatever may have been the solvent action of the rain on the rocks beneath."

XII. CLIMATE.

By the climate of a country is meant the condition of its atmosphere as regards heat or cold, moisture or dryness, healthfulness or unhealthfulness.

The atmosphere, or ocean of air that surrounds the earth, gets practically all its heat from the sun, either directly by absorption as the rays pass through the air, or indirectly from the heated earth.

Or, less concisely, the atmosphere receives its heat from the sun :

1. Directly, by absorption.

2. Indirectly, from the heated earth.

(*a.*) By contact with the heated earth.

(*b.*) By radiation from the heated earth,—that is, the sun's rays heat the earth, and the heated earth throws out or radiates its heat in all directions.

(*c.*) By reflection from the heated earth,—that is, the sun's rays strike the earth and fly off from it like light does when it strikes a mirror.

The equatorial regions of the earth are warmer

than either the temperate or the polar regions, because they receive the sun's rays more directly than any other part of the earth.

Regions of the earth that are situated the same distance from the equator, however, often possess different temperatures, not only because they are exposed to warmer or colder currents of air or water, but also on account of certain peculiarities of their surfaces

The distribution of the land and water areas of the earth exerts a marked influence in causing a difference in climate in regions situated in the same latitude.

A given quantity of the sun's heat falling on a given area of water will produce therein a smaller increase of temperature than if permitted to fall on an equal area of land. Consequently, the air over such body of water will be less warmed than would the air over the land.

Water possesses a greater capacity for heat than any other common substance; in other words, a greater quantity of heat is required to cause a certain increase of temperature in a pound of water than in a pound of any other common substance.

For example: the quantity of heat required to raise a pound of ice-cold water to its boiling-point, or to 212 degrees Fahrenheit, would be sufficient to raise the temperature of a pound of ice-cold iron to about 1600 degrees Fahrenheit, or to make the ice-cold iron red-hot.

Although land and water areas may be situated in the same latitude, and therefore receive equal quantities of the sun's heat per unit of area, yet the temperature of the land, and consequently of the air over it, would become much hotter than the temperature of the water, and of the air over the water.

The higher the temperature of an area, the more rapidly it loses its heat. A land surface, when heating, becomes hotter than a water surface when similarly exposed for the same time to the sun's heat. The land also, when cooling, loses its heat more rapidly than the water; the air over the land becomes chilled sooner than over the water.

Differences in the elevation of the land produce differences in the climate. In general, an elevation of three hundred and fifty feet will cause as great a lowering of temperature as a difference of one degree of latitude, or of about seventy geographical miles. Therefore, the same differences are observed in passing from the base to the summit of a high tropical mountain as are observed in passing along the surface of the earth from the equator to the poles. Or, in other words, three hundred and fifty feet skyward equals seventy miles poleward.

In summer, when the sun is more nearly over-head, and when in our hemisphere the earth is gaining rather than losing heat, the land areas, and consequently the air over them, rapidly become heated; while the water areas, and consequently the air over them, remain comparatively cool.

In winter, however, when the loss of heat is greater than the gain, the land areas, and consequently the air over them, rapidly become cooled; while the water areas continue for a long time to part with the great stores of heat that they have taken in during the summer, and thus remain comparatively warm.

Similar differences are observed between the temperature of the air over the land and water areas during the daylight while they are exposed to the sun's heat, and during the night when they are throwing it off.

There is another reason why the water areas are heated less rapidly than the land areas: the heat penetrates the water to a comparatively great depth, is diffused through a great body of water, and, consequently, heats it less. Moreover, the water when heated is set in motion by reason of the differences of density produced by the differences of temperature, and moves towards colder districts, and its place is taken by water that moves from colder districts. Such motions are seen in the constant ocean currents.

The land, on the contrary, is heated to a comparatively small depth, remains in its place, and may, therefore, rapidly become intensely hot.

The climate produced by an extended land area is called a continental climate. That produced by an extended water area is called an oceanic climate. The continental climate is characterized by great extremes of heat and cold,—that is, a continental climate is apt to be very hot in summer and very cold in winter. The oceanic climate, on the contrary, is characterized by a comparatively uniform temperature, being neither very hot in summer nor very cold in winter.

So far as the climate of the land is concerned, the differences of climate above referred to are greatly influenced by the nature of the surface. If the surface is covered by vegetation of any kind, especially by forests, it both heats slowly and cools slowly.

If entirely bare, or deprived of vegetable covering, it both heats quickly and cools quickly.

The climate of a forest, or, indeed, even of a region protected by a less dense covering of vegetation, closely resembles the equable climate of a water area; that of a bare, arid district, differs greatly from that of a water area. Deserts, for example, are characterized by great extremes of climate, being very warm in summer, or in the daytime, and very cold in winter, or at night.

The climate of a country, therefore, will be greatly influenced by the presence of the forest districts, and must necessarily be changed, to a greater or less extent, by the removal of such forests from extended areas.

Humboldt, in " Cosmos," * on page 318, thus describes the influence of land ·and water areas on oceanic or continental climates; or, as he styles them, on the insular or littoral climates:

"I have already alluded to the slowness with which the great mass of water in the ocean follows the variations of temperature in the atmosphere, and the consequent influence of the sea in equalizing temperatures; it moderates both the

* " Cosmos," vol. i., by Alexander von Humboldt. London: Longman, Brown, Green & Longmans, 1849. Pp. 480.

asperity of winter and the heat of summer: hence arises a second important contrast,—that between insular or littoral climates (enjoyed also in some degree by continents whose outline is broken by peninsulas and bays), and the climate of the interior of great masses of solid land. Leopold von Buch was the first writer who entered fully into the subject of this remarkable contrast, and the varied phenomena resulting from it; its influence on agriculture and vegetation, on the transparency of the atmosphere and the serenity of the sky, on the radiation from the surface, and on the height and limit of perpetual snow. In the interior of the Asiatic continent, Tobolsk, Barnaul on the Obi, and Irkutsk have summers which, in mean temperature, resemble those of Berlin and Munster, and that of Cherbourg in Normandy, and during this season the thermometer sometimes remains for weeks together at 30° and 31° C. (86° or 87.8° F.); but these summers are followed by winters in which the coldest month has the severe mean temperature of 18° to 20° C. (—4° to +4° F.)."

Flammarion, in his work entitled " The Atmosphere," * referring to the contrasts between continental and oceanic climates, says, on page 250 :

"The climate of Ireland, Jersey and Guernsey, of the Peninsula of Brittany, of the coasts of Normandy, and the South of England, countries in which the winters are mild and the

---

* Reprinted, by permission, from "The Atmosphere," by Camille Flammarion. New York : Harper & Brothers, Franklin Square, 1873. Pp. 454.

summers cool, contrast very strikingly with the continental climate of the interior or Eastern Europe. In the northeast of Ireland (54° 56′), in the same latitude as Königsberg, the myrtle grows in the open ground just as it does in Portugal. The temperature of the month of August in Hungary is 69.8°; in Dublin (upon the same isothermal line of 49°) it is 61° at most. The mean temperature of winter descends to 36.3° at Buda. In Dublin, where the annual temperature is only 49°, that of the winter is, nevertheless, 7.7° above the freezing-point, or nearly four degrees higher than at Milan, Pavia, Padua, and all Lombardy, where the mean heat of the year reaches 55°. In the Orkney Islands, at Stromness, a little to the south of Stockholm (there is not one degree difference in their latitudes), the mean winter temperature is 7°, or higher than that of London or Paris. Stranger still, the inland waters of the Faroe Islands never freeze, situated in 62° of north latitude, beneath the mild influences of the west wind and the sea. Upon the coast of Devonshire, one part of which has been termed the Montpellier of the North, because of the mildness of its climate, the Agave Mexicana has been known to flower when planted in the open air, and orange-trees trained upon the wall to bear fruit, though only scantily protected by a thin matting.

"There, as at Penzance, Gosport, Cherbourg, and the coast of Normandy, the mean temperature of the winter is 42°, being but 18.5° below that of Montpellier and Florence."

XIII. CLIMATE AS INFLUENCED BY THE PRESENCE OF THE FOREST.

WHEN sunshine falls on an area covered by trees, the heat is more thoroughly absorbed or taken in than when it falls on an arid or uncovered surface.

The more thorough absorption of heat by a wooded area is caused mainly as follows:

1. The greater extent of surface presented by a wooded than by an unwooded area, not only by the trees themselves, but often also by the under-brush which exists in most forest regions.

2. The porous and better absorbing character of the carpet of leaves that generally covers the ground in forest regions.

3. The presence of a greater amount of moisture in the air of a forest region than in a region that is void of vegetation.

The marked increase in the area of a surface covered by a forest over that of an equally large unwooded surface would itself, apart from any other circumstances, necessitate a smaller rise or in-crease of temperature in the forest than would the same quantity of the sun's heat falling on an equal area of bare ground. Therefore, the sun's heat when permitted to fall on a forest region is more thoroughly absorbed, is spread over a greater surface, and penetrates the ground more thoroughly than it would if thrown on bare ground. For, when the rays fall on a bare, dry, parched surface, they penetrate the ground to but a small depth, and heating a smaller amount, must necessarily produce a greater increase of temperature.

The same is true as regards the loss of heat: forest districts, which take in heat slowly, part with it slowly; while bare, uncovered surfaces, which take in heat quickly, part with it quickly.

It, therefore, follows that since the forests do not rapidly heat, they do not become excessively hot in summer; and, since they part with their heat slowly, they do not become very cold in winter.

The fact that an area covered with forests does not tend to become as cold in winter as bare, uncovered ground, exerts a great influence on the depth to which the frost extends downwards.

The non-conducting power for heat of even a very thin layer of snow is well known. If snow falls before the frost penetrates the ground to any great depth, it will act as a covering to prevent the earth from losing heat. The frost will, therefore, be prevented from entering the ground to any great depth. In the temperate regions of the Northern Hemisphere, in districts covered by forests, the early snows of winter are, for the greater part, apt to fall before the ground is frozen to any considerable depth. In the early spring, when the thaws come, the water derived from the melting of the ice and snow can then drain quietly into the ground and fill the reservoirs of the springs.

If, however, the forests are removed, the ease with which the ground loses its heat generally permits it to freeze before the first snow falls, and the non-conducting power of the layer of snow causes the ground to remain frozen until long after the spring thaws have melted the snow. Under such circumstances, the water derived from the melting snow rapidly drains almost entirely off the surface, and is apt to produce disastrous floods.

Tyndall has shown that the ability of air to become heated, by absorbing heat directly from the sun's rays as they pass through it, depends almost entirely on the presence of water vapor. The air over a forest district is necessarily more moist, and consequently better able to absorb heat and become heated than the air over a dry, barren tract.

i

Moist air, moreover, is not only better able to absorb the heat accompanying the direct rays of the sun, but is especially able to absorb that kind of heat which is thrown off from the heated earth. Consequently, the air over a forest district absorbs a much larger percentage of the heat flung off from the heated earth than does the drier air over a barren district. In this way, in winter, while the ground is throwing off its heat, the moist air of the forest tends to remain warmer than the air over a dry, arid tract.

Forests exert a marked influence on the climate of a country, especially at that time of the year when the crops are liable to suffer injury from early frosts.

The ease with which bare or poorly covered ground throws off its heat permits such an area to more readily reach the temperature of the danger-point than would be the case if it were well wooded. It must be remembered that the difference of a few degrees, or even of the fraction of a degree, between the air over an uncovered district, and the air over a covered district in the forest, may make all the difference between the occurrence of frost and its non-occurrence. It is, indeed, often but the difference of a fraction of a degree, that may cause by an early frost the loss of millions of dollars to an agricultural district.

Again, it is in the late autumn, at the time of the early frosts which are so feared in the agricultural districts, that a vegetable covering may be able to fling back to the earth sufficient heat thrown out by the cooling ground to prevent the temperature of the air immediately around growing plants from reaching the freezing-point.

Forests exert a sheltering action at the time of frosts in keeping the land to the leeward warmer than that to the windward. Not only do they act as an actual barrier or screen, sheltering and protecting the land immediately to the leeward side, but this protecting action extends to a much greater distance beyond the immediate neighborhood of the forest than might be supposed.

The leaves of almost any forest tree, when examined under the microscope, show greatly extended surfaces in the shape of irregularities, or spine-like projections. These extended surfaces aid the tree greatly in throwing off from the very slightly heated earth the stores of heat which it possesses, even in the depth of winter, and which, thus passing into the air, tend to prevent a too marked fall of temperature during winter.

Generally, the air of the forest is cooler and damper in summer than the air over the open fields in the same district.

1st. Because the air of the forest is shielded from the direct rays of the sun.

2d. Because the air is chilled by evaporation from the moister ground.

A large tract of forest in any section of country tends to prevent marked changes in its climate, as compared with those that occur in the same region over the open fields.

1. By permitting the wooded area to more thoroughly absorb the sun's heat on account of the greater surface it presents.

2. By keeping the air moister and, therefore, better able to absorb the sun's heat.

3. By acting as a screen to the land to the leeward of a cold wind.

4. By preventing the frost from penetrating the ground to too great a depth before protected by a covering of snow.

Geikie, in his " Text-Book of Geology," * in

---

* Reprinted, by permission, from "Text-Book of Geology," by Archibald Geikie, LL.D., F.R.S.   London: Macmillan & Co., 1882.   Pp. 971.

referring to the manner in which man may influence the climate of any particular part of the earth, says on page 471:

"Human interference affects meteorological conditions: 1, by removing forests and laying bare to the sun and winds areas which were previously kept cool and damp under trees, or which, lying on the lee side, were protected from tempests; as already stated, it is supposed that the wholesale destruction of the woodlands formerly existing in countries bordering the Mediterranean has been in part the cause of the present desiccation of these districts; 2, by drainage, the effect of this operation being to remove rapidly the discharged rainfall, to raise the temperature of the soil, to lessen the evaporation, and thereby to diminish the rainfall and somewhat increase the general temperature of a country; 3, by the other processes of agriculture, such as the transformation of moor and bog into cultivated land, and the clothing of bare hill-sides with green crops or plantations of coniferous and hard-wood trees."

Not only does the forest prevent the excessive heating of the land on which it grows, and therefore similar excessive heating of the air over the land, by the greatly extended surface the trees and undergrowth present to the sun's rays, but it also acts by the direct absorption of the sun's rays to cause the separation of the carbon from the oxygen in carbonic acid, and the hydrogen from the oxygen in the water in the vegetable kingdom. Tyndall, in his "Heat as a Mode of Motion," * page 529, says:

"In the building of plants, carbonic acid is the material from which the carbon of the plant is derived, while water is the substance from which it obtains its hydrogen. The solar rays wind up the weight. They sever the united atoms, setting the oxygen free, and allowing the carbon and the hydrogen to aggregate in woody fibre. If the sun's rays fall upon a surface of sand, the sand is heated, and finally radiates away as much heat as it receives. Let the same rays fall upon a forest; then the quantity of heat given back is less than that received, for a portion of the sunlight is invested in the building of the trees. We have already seen how heat is consumed in forcing asunder the atoms of bodies, and how it reappears when the attraction of the separated atoms comes again into play. The precise considerations which we then applied to heat, we have now to apply to light, for it is at the expense of the solar light that the chemical decomposition takes place. Without the sun, the reduction of the carbonic acid and water cannot be effected; and, in this act, an amount of solar energy is consumed, exactly equivalent to the molecular work done."

Concerning the influence of the forests on climate, Hough, in his report to the United States

---

* Reprinted, by permission, from "Heat as a Mode of Motion," by John Tyndall, F.R.S., LL.D. New York: D. Appleton & Company, 5 Bond Street, 1883. Pp. 591.

Commissioners of Forestry for 1877, quoting from a paper by M. A. C. Becquerel, on the " Climatic Effects of Forests," * page 310, says:

" The forests exercise in many ways an influence upon the climate, but to understand this we must define what we understand by *climate*.

" The climate of a country, according to M. Humboldt, is the combination of calorific, aqueous, luminous, ærial, electrical, and other phenomena, which fix upon a country a definite meteorological character that may be different from that of another country under the same latitude and with the same geological conditions. According as one or another of these phenomena predominate we call the climate warm, cold, or temperate, dry or humid, calm or windy.

" We always regard heat as exercising the greatest influence, and after this the amount of water falling in different seasons of the year, the humidity or dryness of the air, prevailing winds, number and distribution of storms throughout the year, clearness or cloudiness of the sky, the nature of the soil and vegetation which covers it, and, according as it is natural or the result of cultivation, the following questions arise for consideration:

" 1. What is the part that forests play as a shelter against the winds or as a means of retarding the evaporation of rainwater?

---

* Reprinted, by permission, from a " Report upon Forestry," 1877, by Franklin B. Hough. Washington Printing-Office, 1878. Pp. 650.

" 2. What influence do the forests exert, through the absorption of their roots or the evaporation of their leaves, in modifying the hygrometrical conditions of the surrounding atmosphere?

" 3. How do they modify the temperature of a country?

" 4. Do the forests exercise an influence on the amount of water falling, and upon the distribution of rain throughout the year, as well as upon the regulation of running waters and springs?

" 5. In what manner do they intervene in the preservation of mountains and slopes?

" 6. Do the forests serve to draw from the storm-clouds their electricity, and by thus doing diminish their effects upon the neighboring regions not wooded?

" 7. What is the nature of the influence that they may be able to exercise upon the public health?

" From these questions we may see what questions we must solve before being able to decide as to the influence that the clearing off of woodlands may exercise upon the climate of a country."

Much valuable data concerning the results of the destruction of the forests on climate, rainfall, and other meteorological conditions, were collected by several scientific expeditions sent to Brazil under the direction of Louis Agassiz.

In an account of the Thayer expedition in 1865 and 1866, Professor Hartt, in a description of the " Geology and Physical Geography of Brazil," * page 319, thus refers to the marked effects that have been produced by the wholesale destruction of the forests by the burning over of their former areas,—

"The limits of the forests, of the belt of decomposition, and of the area over which copious rains fall, coincide very remarkably, and show a dependence upon each other, but the forest belt has a smaller area than that of decomposition or of the rains. The wooded belt seems to have narrowed greatly within comparatively recent times, losing its foothold in the west, where immense regions, now campos, over which the climate and soil would normally be proper for the growth of forests, have dried up, the climate has become hot, less rain now falls, and the forest cannot regain its lost place. Doubtless there are many natural physical causes to be taken into consideration in studying the distribution of the forest, catinga, and campos floræ; but there is one agency that has been at work in Brazil whose effects we can hardly over-estimate, and that is the burning over of the wood and campos lands by man. The very physical features of the highlands of Brazil determine a difference in the luxuriance in the floræ of different regions, and there are, as I have already shown, regions where for ages the climate has been such that forests could scarcely have had any noteworthy extension, so that there must have always been in Brazil, naturally, virgin forests, catingas, campos, and barrens. On the coast, where the forest is dense and moist, and the climate is wet, forest fires are next to

---

* Reprinted, by permission, from "Scientific Results of a Journey in Brazil," by Ch. Fred. Hartt. Boston: Field, Osgood & Co., 1870. Pp. 620.

impossible, and one never sees a scorched and dead wood, such as covers so large an area in the province of New Brunswick, for instance. But in the interior, where the catinga forests drop their leaves, and are dead for several months in the dry season, fires are easily kindled and the wood killed; and fires set in open fields or campos, for the purpose of producing a new crop of grass, may spread to the neighboring catingas. It is the opinion of many writers that a large part of the catinga and campos regions of the Brazilian highlands was once covered by forests, and that their present bare appearance and the character of their floræ is in a very great measure due to frequent and extensive burning over of the country. Every year the Brazilian campos lands are systematically and almost entirely burned over, for the purpose of producing a new crop of grass. This burning, of course, has destroyed all those trees and shrubs and plants of all kinds that cannot bear the scorching, and has wrought a great alteration in the character of the whole flora of the region; the climate also has suffered a change, for with the destruction of the woods and forests it becomes hotter, the unprotected earth is like a furnace, streams run dry a few days after a shower, and the springs disappear."

The following geographical instances of the effects of forests on climate are referred to by Becquerel in a previous quotation.*

---

* Reprinted, by permission, from "Report upon Forestry," 1877, by Franklin B. Hough. Washington: Government Printing-Office, 1878.

St. Helena.

Fully forested when discovered in 1502. The introduction of goats and other causes led to the removal of its forests. Heavy floods and severe droughts were the result; replanting of forest trees towards the close of 1700 resulted in a more uniform rainfall and its better distribution. Subsequent destructions of the forest have again brought back the original condition of affairs.

Island of Ascension.

When discovered in 1815 it was barren, and so destitute of water that supplies were brought to it from the mainland. The effects of planting trees resulted in an increased rainfall, from 10.18 in 1858 to 25.11 in 1863. It now grows forty kinds of trees, where but one grew in 1843 for want of water.

XIV. PURIFICATION OF THE ATMOS-PHERE.

THE atmosphere covers the earth's surface as a vast ocean of air that extends upwards for a distance of several hundred miles.

It is composed mainly of a mixture of two gaseous substances,—namely, of nearly seventy-seven per cent. by weight of nitrogen and about twenty-three per cent. of oxygen. Besides these there is a nearly constant quantity of carbonic acid gas, and a variable quantity of the vapor of water.

The carbonic acid is nearly in the proportion of four parts to ten thousand of air; or very nearly one cubic inch of carbonic acid gas to each cubic foot of ordinary air.

The different ingredients of the atmosphere serve various purposes in the economy of the earth.

The oxygen is necessary for the existence of animal life.

The carbonic acid is necessary for the existence of plant life.

The moisture of the air is necessary for the existence of both animal and plant life, although, perhaps, it is more necessary for the existence of plant life.

Every action of an animal results in the decay and subsequent death of some part of its body. Although this death does not take place immediately, yet the use of any part or member of the body results in its waste and subsequent death. In order to replace these dead parts some form of nourishment is necessary. This nourishment comes from the food of the animal, which, by the process of digestion, goes to make up the blood. The blood carries to the parts of the body which require nourishment the materials needed for subsequent growth, and, at the same time, takes away or carries off the dead or decaying parts.

The blood is forced through the different parts of the body by the action of the heart, which acts like a force-pump. The blood goes to these parts of the body as bright red arterial blood. It leaves them so clogged with dead and decaying parts, that it becomes changed into a dark, bluish-black, venous blood.

The oxygen of the air is, in general, necessary to the existence of animal life in order to burn out or remove from the blood these dead and decaying parts, and so change the dark, venous blood to bright red arterial blood.

The oxygen brings about this change mainly by combining with and slowly burning the waste products so as to form water vapor and carbonic acid gas.

If there was nothing to oppose this action of animal life all the oxygen would, in the end, be removed from the air and changed into carbonic acid gas, and no further animal life would be possible on the earth. Plants, however, during their growth, in the presence of sunshine, take in or absorb carbonic acid gas. In the delicate structure of the leaf this gas is broken up into carbon, which is retained by the plant to form its woody fibre, and into oxygen, which is given off and passes into the atmosphere.

Plants, therefore, during active growth take in carbonic acid gas and give out oxygen.

A wonderful balance is thus maintained in nature, and the composition of the atmosphere is kept practically constant.

What animals reject, plants need for their existence. What plants reject, animals need for their existence. It is like the case of the renowned

Jack Sprat and his wife, of nursery lore, who between them kept both plate and platter clean. Each liked and thrived on what the other rejected.

If this balance between the plant and animal kingdom is disturbed, the composition of the atmosphere will be altered, and a marked change will be produced in the earth's plant and animal life. Such changes have been observed in the geological past long before the creation of man.

The earth's atmosphere was originally vaster than at present. The quantity of oxygen and carbonic acid gas was enormously greater.

A careful estimate places the amount of oxygen that exists, combined with the different substances that form the fifteen or twenty miles of the earth's crust that have been carefully studied, at least at one-half of the total weight.

At an early age in the earth's life this oxygen existed in a free state in the atmosphere, and became fixed by combining with or oxidizing the different materials of the crust. This oxidizing action has now practically ceased, and the quantity of oxygen present in the air is constant.

Prior to the Carboniferous age carbonic acid

must have existed in the air in enormous quantities; for, the vast deposits of carbon, which form the coal-beds now found in the different parts of the crust, then existed in a gaseous condition in the atmosphere combined with oxygen.

Animal life of the present type was impossible in the unpurified atmosphere that existed before the Carboniferous age. The conditions were, however, such as to favor dense and luxuriant plant life, and at no time in the world's history, either before or since, has such luxuriant vegetable growth existed.

A twofold action of purification was effected by the plants of the Carboniferous period; namely, the separation of the carbon and the liberation of the oxygen.

The exact balance between the plant and animal life of the earth, so carefully established by nature, cannot be disturbed without marked changes in the entire races of animals and plants that now exist.

The thoughtless and unnecessary removal of the forests from over extended areas will not only disturb the balance during the time such surfaces are bare, but since, in many cases, such removal permits this section of country to be denuded of its soil, there will also follow a permanent disturbance from the inability of such section of country to sustain any plant life.

As to the purification of the atmosphere by plants, Dana, in his "Manual of Geology," * says, on page 353,—

"In the present era, the atmosphere consists essentially of oxygen and nitrogen, in the proportion of twenty-three to seventy-seven parts by volume. Along with these constituents, there are about four parts by volume of carbonic acid in ten thousand parts of air. Much more carbonic acid would be injurious to animal life. To vegetable life, on the contrary, it would be, within certain limits, promotive to growth; for plants live mainly by means of the carbonic acid they receive through their leaves. The carbon they contain comes principally from the air.

"This being so, it follows, as has been well argued, that the carbon which is now coal, and was once in plants of different kinds, has come from the atmosphere, and, therefore, that the atmosphere now contains less carbonic acid than it did at the beginning of the Carboniferous period, by the amount stowed away in the coal of the globe.

*   *   *   *   *   *   *   *

"Such an atmosphere, containing an excess of carbonic acid as well as of moisture, would have had greater density than

---

* Reprinted, by permission, from a "Manual of Geology," third edition, by James D. Dana. New York: Ivison, Blakeman, Taylor & Co. London: Trübner & Co. Pp. 911.

the present; consequently, as urged by E. B. Hunt, it would have occasioned increased heat at the earth's surface, and this would have been one cause of a higher temperature over the globe than the present.

"During the progress of the Carboniferous period there was, then, (1) a using up and storing away of the carbon of the superfluous carbonic acid, and, thereby, (2) a more or less perfect purification of the atmosphere, and a diminution of its density. In early time there was no aerial life on the earth; and, so late as the Carboniferous period, there were only reptiles, myriapods, spiders, insects, and pulmonate mollusks. The cold-blooded reptiles of low order of vital activity, correspond with these conditions of the atmosphere. The after-ages show an increasing elevation of grade and variety of type in the living species of the land."

XV. HAIL.

HAIL occurs at times when great differences of temperature exist between neighboring masses of very moist air.

By permitting great differences of temperature to occur, the destruction of the forest is, in many cases, followed by an increase in the number and severity of hail-storms.

In order to understand the manner in which the destruction of the forest may influence the occurrence of hail-storms, it will be necessary to study some of the peculiarities of such storms and to review what are now generally believed to be their causes.

Although hail may fall at any time of the year, yet it occurs most frequently in summer towards the close of a very warm day.

The exact causes which produce hail are not known. The conditions necessary for its occurrence appear to be the rapid mixture of very warm and very cold moist air.

A hail-storm is generally preceded by the appearance of several layers of dark, grayish clouds. In most all cases, before the beginning of a hail-storm, a violent movement is seen to take place between these layers, apparently of a whirling character. Generally, too, hail-storms are attended by violent disturbances in the electrical equilibrium of the atmosphere, as is evidenced by the frequent discharge of the lightning-bolt and the almost continual roar of thunder. Then follows a fall of hailstones, the size of which is much larger at the beginning and towards the middle of the storm than towards the close. Towards the close of the storm, however, the quantity of hail which falls is greatest.

If a hailstone be examined by cutting it in two, it will be seen to consist of alternate layers of ice and snow laid over one another in successive coats like the layers of an onion. A cross-section of a hailstone can be made by holding it against the surface of a hot plate until half of the stone has been melted away.

Hailstones vary in weight from a few grains to several pounds. Records exist of hailstones weighing many pounds, sometimes of even several hundred pounds. In such cases, however, it is more than probable that the stones were produced by the regelation, or freezing together, of numerous smaller stones, as follows:

The excessive fall of small hailstones, that occurs towards the close of the storm, often produces heaps of hailstones several feet in thickness. The separate hailstones readily freeze together, and are afterwards cut into smaller masses by the action of the water rapidly draining off the earth. The fragments thus formed, in all probability, give rise to stories of mammoth hailstones.

The severity of the lightning-flashes, which attend nearly all great hail-storms, has led some meteorologists to believe that hail-storms are caused by the presence of an unusual quantity of free electricity in the atmosphere. The electrical theory of hail is, however, at the present, almost entirely discarded, it being now generally recognized that the lightning is the effect of the hail-storm, and not its cause.

Volta proposed the following electrical theory for the production of hail. He imagined two approximately parallel clouds near together, the upper cloud formed of snow, and the under cloud of rain. Assuming these clouds to be respectively charged with positive and negative electricity, the particles of snow in the snow-cloud might, he assumed, be alternately attracted and repelled into and from the rain-cloud, and thus receive alternate coatings of ice and snow until they finally fell to the ground as hailstones.

In France, where the reckless destruction of the forest has been attended by a marked increase in the number and severity of hail-storms, miniature lightning-rods have been erected in the fields to prevent the occurrence of hail-storms. These lightning-rods either took the shape of captive balloons secured to the earth by tinsel threads, or of bundles of straw set upright in the field, or of metal rods permanently connected with the ground. Their object was to gradually discharge the air of its free electricity, and thus prevent the occurrence of hail-storms. The name of such rods, paragrêles, is significant of their supposed action. Unfortunately, they have proved futile in action, since again and again the portions provided with this supposed protection have been as severely visited by hail-storms as unprotected portions.

An endeavor has been made to explain the peculiar shape of hailstones by the existence of a number of approximately parallel clouds composed alternately of snow and rain. Drops of rain falling from the upper cloud would thus receive successive coatings of snow and ice as they passed successively through the snow- and rain-clouds, and would finally fall as characteristically-shaped hailstones.

The theory now generally received in regard to the formation of hailstones is, that in such storms the wind rotates around a vertical rather than around a horizontal axis. If such a whirling motion exists between a neighboring rain- and snow-cloud, the particles of snow would be successively dipped into the rain- and snow-clouds, and would thus receive alternate layers of ice and snow.

A somewhat similar theory regards a hail-storm as belonging to the type of the ordinary tornado. The whirling motion of the air is supposed to produce the alternate coatings of ice and snow by the alternate exposure of the moisture to the different temperatures found in the denser and rarer portions of the space around which the wind is whirling.

Hail-storms often cause great damage. A single hail-storm in France has been known to cause loss to the agricultural districts amounting to the sum of at least one million pounds sterling.

Although the exact cause of hail-storms is at present unknown, yet the storms never occur

unless marked differences of temperature exist
between neighboring portions of the air.   The
removal of the forest from any considerable section
of country permits such differences of temperature to occur.   In point of fact, it has been noticed in parts of the world from which the forests
have been removed, that the number and severity
of hail-storms have undoubtedly increased.

Destructive hail-storms may therefore be regarded as one of the evil results which naturally
follow the destruction of the forest.

As regards the supposed protective influence
of lightning- or hail-rods against destructive hail-
storms, Loomis, in his " Treatise on Meteorology," *
writes on page 135 :

"It has been proposed to preserve the vineyards and valuable farms from the ravages of hail by erecting an immense
number of poles, armed with iron points, communicating with
the earth, for the purpose of drawing off the electricity of the
clouds.   Multitudes of these hail-rods were erected in Switzerland, but without the expected success.

"It is believed that electricity performs altogether a subordi-
nate, if not an unimportant part in the formation of hail;

---

* Reprinted, by permission, from " A Treatise on Meteorology," by Elias Loomis, LL.D.   New York: Harper &
Brothers, Franklin Square, 1868.   Pp. 305.

and if we could draw off all the electricity from the hail-cloud as fast as it was generated, it is not improbable that the hail would be formed about as large and as abundantly as at present.

"But, even supposing electricity to be the sole agent in the production of hail, hail-rods could not be expected to furnish security against hail unless an entire continent could be studded thick with them, for in the middle latitudes the hail-cloud advances eastward with a velocity sometimes of forty or more miles per hour, and the hailstones which fall in one locality are those which were forming when the cloud was many miles westward of that point; so that, to protect a small spot, the whole country for many miles westward should be armed with rods; and it is conceivable that a hail-cloud arriving over a region studded with these rods might immediately pour down a large quantity of hailstones which would have fallen farther eastward if the rods had not discharged the electricity of the cloud."

The following description of a hail-storm that occurred near Bordeaux, France, in 1865, is thus given by Flammarion, in his work entitled "The Atmosphere," * page 393.

"On May 9, 1865, for instance, a storm began at 8.30 A.M. over Bordeaux and proceeded in a N.N.E. direction, passing over Périgueux at 10 A.M., Limoges at noon, Bourges at 2 P.M., Orléans at 5.30 P.M., Paris at 7.45 P.M., Laon at 11 P.M., and collapsing a little after midnight in Belgium and the North

---

* Reprinted, by permission, from "The Atmosphere," by Chamille Flammarion. New York: Harper & Brothers, Franklin Square, 1873. Pp. 454.

Sea. Its mean breadth was from fifteen to twenty leagues. The hail only fell in certain places : to the left of Périgueux, over the arrondissement of Limoges, to the right of Châteauroux, to the southeast of Paris, from Corbeil to Lagny, and in the arrondissements of Soissons and Saint-Quentin. At this latter point it was of a formidable character. The crystal mass which fell from the sky upon the Catelet meadows formed a bed a mile and a quarter long and two thousand feet broad, estimated to amount altogether to twenty-one millions of cubic feet. The hailstones did not disappear for more than four days afterwards. These hailstones sometimes destroy all the crops, as, for instance, that which occurred in the neighborhood of Angoulême on August 3, 1813. The day had been fine, and the wind was due north until 3 P.M., when it suddenly veered right round ; the sky gradually became covered with clouds, which, collecting one on the top of the other, offered a terrible spectacle. The wind, which from noon until 5 P.M. had been rather violent, suddenly dropped. Thunder was heard in the distance, and gradually became louder ; the sky, at last, became totally obscured, and at 6 P.M. there was a tremendous fall of hail, the stones being as large as eggs. Several persons were severely wounded, and a child was killed near Barbezieux. The next day the ground looked as it might do in midwinter : the hailstones had accumulated in the hollows and the roads to a height of thirty to forty inches ; trees were entirely stripped of their leaves ; vines were cut into pieces, the crops crushed, the cattle, sheep, and pigs especially were severely injured. The whole neighborhood was deprived of game, and some few young wolves were found dead. The effects of the storm were still visible in 1818."

XVI. REFORESTATION AND TREE-PLANTING.

By reforestation is meant the replanting of trees in any locality from which they have been removed either accidentally or purposely.

Where the removal of the forests has been made in a hap-hazard way, and no care has been taken to protect the soil from the effects of rapid drainage, the loss of the soil in some cases is so great as to render the area not only unable to sustain trees of the same character as those which have been removed, but even to render it unable to sustain any trees whatever. If, however, in the removal of the trees, care has been taken that the loss of soil is but trifling, it may be possible to again reclothe the surface with trees similar to those which have been removed.

The French term for this process of replacing forests is *reboisement.* Our English word reforestation may be safely taken as its equivalent.

The object of reforestation is to avoid the evils which result from the removal of the forests by

perpetually maintaining forest tracts in portions
of the earth set aside for such purposes.

Where the loss of the soil following the destruction of the forests has been too marked to permit
the successful replanting of trees, some of the evil
effects following rapid drainage, such, for example, as disastrous floods, with their consequent
droughts, have been in a measure lessened by replanting the bare surface with different species of
hardy grasses, which, by absorbing and holding
the rain, permit the water to drain slowly off the
surface.

The time required for the full growth of forest
trees is so great that, unless considerable encouragement is given to the planting of trees, reforestation will scarcely be attempted to any considerable
extent. In most cases where reforestation has been
attempted, laws have been enacted offering certain
premiums, either in land or in money, for successful tree-planting.

Where reforestation is carried out on a large
scale, under the encouragement of a government, it
is desirable that either seeds or seedlings be supplied by the government, or that extensive nurseries be established. Great care must be taken
to insure the planting of the varieties of trees best

suited to exist in the particular section of country that is to be reforested.

Since those sections of country where reforestation is to be attempted have already, by the removal of the forests, been exposed to the loss of soil, great care must be taken in the replanting of trees not to needlessly disturb the soil. Two methods may be employed in reforestation,—viz.:

1. Sowing.
2. Tree-planting.

Considerable difference of opinion exists as to which of these two methods is preferable. Unquestionably, however, each is best suited for particular cases, and, in point of fact, each has been adopted with considerable success in different parts of the world.

Seeding can, perhaps, be most profitably followed in the temperate latitudes, in situations where the growth of the tree is comparatively certain. In higher latitudes the planting of trees is, perhaps, preferable, since the germination and continued growth of the seeds are by no means so certain.

In the case of the destruction of the forest by avalanches, replanting or reforestation is rendered much more difficult by the fact that the soil, in such cases, is often so almost entirely removed by the force of the rushing snow that little but the bare rocks remain.

In Italy, laws passed in 1877, set aside the following classes of lands as suitable for being included under the provisions of the forest regulations,—namely:

Forest lands on mountain-sides, or in such places as might, from their location, by the loss of their trees, cause injury to the lowlands by avalanches, or that might, by their drainage, influence or modify the water-courses.

It is generally recognized in some of the western portions of the United States, that when trees are planted in plots around the farm lands, or on the sides of such lands, the protection thus afforded the rest of the land against the winds is greater in actual money value than the rent of the ground occupied by such forests.

Whenever reforestation is attempted over extended areas, care should be taken as to the portions which are best suited for such purposes.

It would seem that the following locations are especially adapted as being suited for the perpetual maintenance of forests on them,—namely:

1. Wet lands.

2. Lands covered with thin soil.

3. Lands covered with a rather poor soil.

4. Along the margins of all rivers, wherever the land is not actually required for purposes of roads or other public uses.

5. On the sides of all mountain-slopes, where the soil is of the proper character.

6. On the slopes of all mountains subject to avalanches.

The following translation of the French laws of 1860 for reforestation is taken from the report of the United States Commissioners for 1877,* page 338:

FRENCH CODE OF REBOISEMENT OF MOUNTAINS, JULY 28, 1860.

"ARTICLE 1. Subventions may be allowed to communes and public bodies, or to individuals, for replanting lands on the tops or slopes of mountains.

"ARTICLE 2. These aids may consist either in the delivery of seeds, or plants, or in premiums in money. In those given by reason of the work done for the general good, and in cases of communes and public bodies, regard is to be had to their resources, and the sacrifices they must make, and to their need, as also to the sums given by general councils for reboisement.

---

* Reprinted, by permission, from a "Report upon Forestry," 1876, by Franklin B. Hough. Washington: Government Printing-Office, 1878. Pp. 649.

"ARTICLE 3. Premiums in money given to individuals cannot be paid until after the work is done.

ARTICLE 4. In cases where the public interests demand that the works of reboisement should be made obligatory, either on account of the condition of the soil, or the dangers that may happen to the lands below, proceedings are to be had as follows :

"ARTICLE 5. An imperial decree, issued in council of state, declares the public utility of the works, fixes the boundaries of land in which it is necessary to execute the reforesting, and the time within which it must be done. This decree is preceded (1) by an open inquiry in each of the communes interested; (2) by a deliberation in the municipal councils of these communes, in conjunction with those most important; (3) the advice of a special commission, composed of the prefect of the department or his delegate, a member of the general council, a member of the council of arrondissement, an engineer of bridges and roads or of mines, a forest-agent, and two landholders of the commune interested; (4) the advice of the council of arrondissement, and that of the general council.

"The *procès-verbal* specifying the lands, the plan of the places, and the project of the works prepared by the forest administration, with the concurrence of an engineer of bridges, roads, or of mines, are to be deposited in the office of the mayor during the inquiry, the duration of which is one month, beginning with the publication of the prefectoral order, which prescribes the opening of the inquest and the meeting of the municipal council.

"ARTICLE 6. The imperial decree is to be published and posted up in the communes interested. The prefect is also to notify the communes and public bodies, as well as individuals, by an extract of the imperial decree, concerning the indications relating to the lands belonging to them. The act of notification shall show the limit of time allowed for the work of reboisement, and if there is occasion, the offer of aid from the administration on the advances it is disposed to make.

"ARTICLE 7. If the lands included within the limits fixed by the imperial decree belong to individuals, the latter are to declare whether they will undertake to do the replanting themselves; and, if so, they are to be held to execute the work within the time fixed by the decree. In case they refuse or fail to perform agreement, proceedings may be had for their expropriation, on the ground of public utility, observing the formalities prescribed under Title II. and following, of the law of May 3, 1841. The proprietor expropriated in the execution of this article has the right to regain possession of his property after reboisement, subject to payment of charges for expropriation, the cost of labors in principal and interest. He may relieve himself of the price of the labors by relinquishing half of the property. If the proprietor wishes to obtain repossession, he should make a declaration to the sub-prefect within five years after notice that the work of reboisement has been finished, under penalty of forfeiture of this right.

"ARTICLE 8. If the communes or public bodies refuse to execute these labors upon their lands, or if they are unable to do it, the state may acquire, either amicably obtaining a part of the lands which they will not or cannot replant, or by assuming sole charge of the work. In the latter case, it will retain the care and use of the lands until it is reimbursed its

l   14*

advances, in principal and interest. Nevertheless, the commune shall enjoy the right to pasturage on the lands replanted as soon as it is found beyond risk of injury.

" ARTICLE 9. Communes and public bodies may in all cases exonerate themselves from repayment to the state, by relinquishing one-half of the replanted lands. This abandonment should be made under loss of right of doing so, within ten years from notice of the completion of the works.

" ARTICLE 10. The sowing or planting cannot be made on more than a twentieth in one year of the surface to be planted, unless a resolution of the municipal councils authorizes it to be done to a greater extent.

" ARTICLE 11. Forest-guards of the state may be appointed for the care of the sowing or planting done within the boundaries fixed by imperial decrees. Injuries proved by these guards, within the extent of these limits, shall be prosecuted in the same manner as if done in woods subject to forest regulation. The execution of the sentence is to be in accordance with articles 209, 211, and 212, and paragraphs 1 and 2 of article 210 of the Forest Code.

" ARTICLE 12. Paragraph 1 of article 224 of the Forest Code is not applicable to reboisement done with aid or premiums from the state, in execution of the present law. The owners of lands replanted with aid or premiums of the state may not pasture their cattle without special license from the forest administration, until the time when such woods shall be recognized by said administration as sufficiently protected.

" ARTICLE 13. A regulation of the public administration shall determine (1) the measures to be taken for fixing the boundaries indicated in article 5 of the present law; (2) the rules to be observed in preservation of works of reboisement; (3) the mode of determining the advances made by the state, and the measures proper for assuring repayment of principal and interest, and the rules to be followed in the relinquishment of lands which article 9 allows communes to make to the state.

"ARTICLE 14. The sum of ten million francs is appropriated for paying the expenses authorized by the present law, to the extent of one million a year. The minister of finances is authorized to sell, with right of clearing, if necessary, woods belonging to the state, to the value of five million francs.

"These woods may only be taken from such as are entered in Table B, appended to this law. The sales shall be done in succession, within ten years from January 1, 1861. The minister of finances is likewise authorized to sell to communes, upon approved valuation, and on conditions fixed by a rule of the public administration, the woods hereinabove mentioned. The five million francs needed to complete the expenses authorized by the present law shall be provided by means of extraordinary cuttings, and, if necessary, from the ordinary resources of the budget."

XVII. THE BALANCE OF NATURE.

FOR such a complex organization as the earth to be properly maintained in operation, an exact balance must be preserved between its five great geographical forms or parts,—namely, the land, the water, the air, plants, and animals. So intimately are these different parts associated with one another, and so exact is the balance that is maintained between them, that no one can be changed, either in amount or distribution, without markedly affecting all the others.

The five geographical forms receive practically, entirely from the sun, all the energy by which they are actuated, and which activity constitutes the order of created things.

A part of the heat of the sun stirs the air or water into vast movements called currents that flow between the equator and the poles. By their means an interchange is effected between the excessive heat of the equatorial regions and the excessive cold of the poles. Another part heats the earth's surface, and causes vapor to pass off from the water surfaces into the atmosphere.

Another part of the solar energy or heat is directly expended in maintaining one or another of the myriad forms of plant and animal life.

If any of the five great geographical forms appropriates more than its share of the solar energy, a disturbance of the balance of nature is effected, which may produce far-reaching changes in the operation of the entire mechanism.

Let us inquire as to some of the more evident ways in which this balance of nature is preserved, how it may be disturbed, and some of the effects produced by such disturbances.

We will discuss this influence from the standpoint of the five great geographical forms,—namely, the land, the water, the air, plants, and animals.

The Land and Water.—An exact balance, both in the amount and distribution, of the land and water areas of the earth is absolutely necessary for the existence of the earth's present plant and animal life.

The total water areas of the earth are in excess of the land areas in about the proportion of $2\frac{1}{3}$ to 1.

The most extended water areas are situated in the equatorial regions, the greater part of the land areas being situated either in the temperate or in the polar zones.

At the equator, therefore, where the sun's heat is greatest, there exists the greatest expanse of water. Here are three readily-movable elements, the air, the water, and vapor, each of which can take in considerable heat without growing very hot. The differences between the temperature of the equatorial and polar regions produce vast currents, both in the atmosphere and in the ocean, which effect an interchange between the excessive heat at the equator and excessive cold at the poles.

Even a comparatively small change in the distribution of the land and water areas of the earth would produce marked changes in its life.

If, for example, most of the earth's surface in the equatorial regions was composed of land, an excessive temperature would be thereby produced that would render the equatorial regions absolutely uninhabitable by any of the present races of man. Consider, for example, tropical Africa. The equator by no means crosses this continent at its greatest breadth, and yet, notwithstanding the fact that nearly all the continent is considerably more than one thousand feet above the sea level, large parts of its interior are as yet absolutely unknown to the white man.

What, then, would be the effect on the earth's present life if, instead of the present excess of water surface at the equator, there existed an excess of land surface? Beyond doubt the present life of the earth would be swept out of existence.

In the same manner any marked increase in either the elevation or the extent of the land in the polar regions would be followed by such an increase in the severity of the cold as to sweep out of existence much of the present life of the earth. It was, in the opinion of most geologists, a change in the elevation of the polar lands that caused the severe cold of the glacial epoch, when most of the northern continents were covered with enormous ice-fields.

The Air.—Any change in the composition of the earth's atmosphere, such, for example, as in the amount of its oxygen or its carbonic acid gas, would be followed by a change in its animal and plant life.

The existence of animal life tends to decrease the amount of oxygen in the atmosphere, and to increase the amount of carbonic acid. The existence of plant life tends to increase the amount of oxygen and to decrease the amount of carbonic acid gas.

A wonderful balance is maintained in nature as to the composition of the atmosphere, from the fact that what plants reject, animals require for their existence, and what the animals reject, plants require.

Minerals, Plants, and Animals.—The mutual interdependence of the mineral, the plant, and the animal affords another illustration of the balance of nature. Animals obtain their food either from other animals or from plants. Plants, as a rule, live on minerals. They are so constituted as to be able to take the various substances directly from the soil, and to change them into forms that can be readily assimilated by animals. The continued existence of animals depends on the continued existence of plants.

Nature has very carefully insured the presence of those germs or seeds that are absolutely necessary for the birth of either animals or plants. To insure the presence of the germs in all cases, the number of such germs produced is always vastly in excess of the number that can possibly live. In the case of nearly all plants and animals the number of germs produced by a single individual is so great, that if they all lived and reproduced their kind at the same rate, in a very little while the earth itself would be too small to hold them.

Leuwenhoek has calculated that a single specimen of the domestic fly can produce seven hundred and forty-six thousand four hundred and ninety-six young in three months.

According to Professor Owen, a single aphis, or plant louse, in the tenth generation produces one quintillion young.

It has been calculated that if all the offspring of a single edible oyster survived for but a comparatively few generations, the waters of such shallow inlets of the ocean as the Chesapeake Bay would be too small to hold them.

In order to avoid this excessive multiplication of the animal and plant life of the earth,—and the above are but a few of the numerous similar cases that might be quoted,—and thus preserve the balance of nature, which would be disturbed by such inordinate multiplication of any one species, all forms of animate creation have their natural enemies provided by nature to hold them in check. Those only continue to exist that are best fitted to exist under the conditions by which they are surrounded.

The principle of the survival of the fittest plays an important part in preserving the balance of nature.

Nearly every animal forms the food best fitted to sustain the life of some other animal. In the event of a too rapid multiplication of any particular form of life, some scourge or disease appears which sweeps off the surplus and thus restores nature's balance.

As far as careful measurements have been made, it can be safely assumed that the total value of the solar radiation is practically the same now as it was many thousands of years ago. Consequently, the total amount of energy which the earth thus receives from the sun, and which goes to maintain the present mechanism of nature, is constant.

The distribution of this solar energy is, however, by no means constant. The general interchange that is effected between the excessive heat of the equator and the excessive cold of the polar regions may take place rapidly or slowly, and thus produce differences in the earth's general climate that not infrequently give rise to a belief in a change in the total heating power of the sun, when no such change exists. For example:

A bare, uncovered surface heats with extreme rapidity, and consequently the air over it becomes intensely heated. This may give rise to an impression that the sun's heating power has increased.

Certain causes may tend to temporarily prevent the free interchange of heat energy that usually exists between hot and cold parts of the earth. There will thus result an increase of temperature in one locality and a marked deficit in another, which would thus give rise to the impression that variations in the solar radiation existed, when, in reality, such variations existed.

In the case of the evaporation of water effected by the sun, if the total value of the sun's heat be constant it might at first sight be supposed that the total quantity of evaporation must remain constant, and that, therefore, the total quantity of heat remaining the same, no change in its distribution could effect a change in the amount of the evaporation, and, consequently, in the value of the rainfall. It must be remembered, however, in this connection, that if circumstances existed in the air of any locality by which during the time of greatest heat the moisture was retained in the air of such locality, and not be removed therefrom, evaporation would necessarily be much smaller than if such moisture were removed by any cause.

The total quantity of the evaporation, therefore, would by no means be constant.

It is possible, therefore, that while the existence of the forest over extended sections of country tends on the whole rather to vary the distribution of the rainfall through a change in the rapidity of the drainage, that, nevertheless, it may also, to some extent, tend to produce a change in the total quantity of the rainfall.

The exact balance of nature that is required to be maintained, in order that the present life of the earth shall exist, can be disturbed by many means. In perhaps no other way does man tend more to disturb this balance than by the destruction of the forests. The removal of the forests from over extended areas effects a disturbance of the balance of nature that is manifested in the following ways :

1. By a marked change in the heat in summer and the cold of winter in the regions formerly covered by forests.

2. By a marked change in the average amount of moisture present in the atmosphere over such regions.

3. By a marked change in the character of the soil in such region.

4. By a marked change in the drainage of such region.

5. By a marked change in the number and severity of floods and droughts in such regions.

6. By a marked change in the salubrity of the regions through which the rivers flow which rise in such districts.

7. By a marked change in the number and severity of hail-storms in such regions.

8. By an increase in the damage to the agricultural districts arising from the appearance of early frosts in or near such regions.

The preservation of the forests, in at least certain localities, is, therefore, imperatively demanded in order to maintain the general balance of nature, and to insure on the earth a place for the comfortable habitation of man.

George P. Marsh, in his work entitled "The Earth as Modified by Human Action," * says on page 8 :

"The revolutions of the seasons, with their alternations of temperature and of length of day and night, the climates of different zones, and the general conditions and movements of

---

* Reprinted, by permission, from "The Earth as Modified by Human Action," by George P. Marsh. New York: Scribner, Armstrong & Co., No. 654 Broadway, 1874. Pp. 656.

the atmosphere and the seas, depend upon causes for the most part cosmical, and, of course, wholly beyond our control. The elevation, configuration, and composition of the great masses of terrestrial surface, and the relative extent and distribution of land and water, are determined by geological influences equally remote from our jurisdiction. It would hence seem that the physical adaptation of different portions of the earth to the use and enjoyment of man is a matter so strictly belonging to mightier than human powers, that we can only accept geographical nature as we find her, and be content with such soils and such skies as she spontaneously offers.

" But it is certain that man has reacted upon organized and inorganic nature, and thereby modified, if not determined, the material structure of his earthly home. The measure of that, reaction manifestly constitutes a very important element in the appreciation of the relations between mind and matter, as well as in the discussion of many purely physical problems. But, though the subject has been incidentally touched upon by many geographers, and treated with much fulness of detail in regard to certain limited fields of human effort and to certain specific effects of human action, it has not, as a whole, so far as I know, been made a matter of special observation, or of historical research, by any scientific inquirer. Indeed, until the influence of geographical conditions upon human life was recognized as a distinct branch of philosophical investigation, there was no motive for the pursuit of such speculations ; and it was desirable to inquire how far we have, or can, become the architects of our own abiding-place, only when it was known by the mode of our physical, moral, and intellectual being is affected by the character of the home which Providence has appointed, and we have fashioned, for our material habitation.

*　　*　　*　　*　　*　　*　　*　　*

"We cannot always distinguish between the results of man's action and the effects of purely geological or cosmical causes. The destruction of the forests, the drainage of lakes and marshes, and the operations of rural husbandry and industrial art have unquestionably tended to produce great changes in the hygrometric, thermometric, electric, and chemical condition of the atmosphere, though we are not yet able to measure the force of the different elements of disturbance, or to say how far they have been neutralized by each other, or by still obscurer influences; and it is equally certain that the myriad forms of animal and vegetable life which covered the earth when man first entered upon the theatre of a nature whose harmonies he was destined to derange have been, through his interference, greatly changed in numerical proportion, sometimes much modified in form and product, and sometimes entirely extirpated.

"The physical revolutions thus wrought by man have not, indeed, all been destructive to human interests, and the heaviest blows he has inflicted upon nature have not been wholly without their compensations. Soils to which no nutritious vegetable was indigenous; countries which once brought forth but the fewest products suited for the sustenance and comfort of man—while the severity of their climates created and stimulated the greatest numbers and the most imperious urgency of physical wants—surfaces the most rugged and intractable, and least blessed with natural facilities of communication, have been brought in modern times to yield and distribute all that supplies the material necessities, all that contributes to the sensuous enjoyments and conveniences, of civilized life. The Scythia, the Thule, the Britain, the Germany, and the Gaul which the Roman writers describe in such forbidding terms have been brought almost to rival the native luxuriance and easily-won plenty of Southern Italy; and, while the fountains of oil and wine that refreshed old Greece and Syria and Northern Africa have almost ceased to flow, and the soils of those fair lands are turned to thirsty and inhospitable deserts, the hyperborean regions of Europe have learned to conquer, or rather compensate, the rigors of climate, and have attained to a material wealth and variety of product that, with all their natural advantages, the granaries of the ancient world can hardly be said to have enjoyed."

XVIII. PRIMER OF PRIMERS.

FORESTRY treats of the care and preservation of parts of the earth covered by trees, together with the best means of replanting such areas when deprived of their trees.

When the germs are present, trees will grow naturally wherever suitable conditions of soil, heat, and moisture exist.

The climatic conditions best suited for the growth of trees are also best suited for the growth of men. As density of population increases, the trees must be removed from large areas :

1. For agricultural purposes.
2. For the location of roads.
3. For the wood or other products.

The principal product of the forest is wood, which is required for fuel or charcoal, for building purposes generally, for fences, for telegraph-poles, for mining purposes, for railroad ties, or for bark for tanning.

The object of forestry is to regulate the removal of the forest where necessary, and to point out the

m

best manner in which the products of the forest may be harvested.

Forestry does not endeavor to preserve intact the virgin forests of the earth. On the contrary, it teaches man how best to harvest the crops of wood, or, where necessary, to safely effect the entire removal of the forests.

Among the different kinds of areas in agricultural districts suitable for tree-planting are,—

1. Areas covered with poor or thin soils, where other crops will not thrive.

2. Wet places, where other crops will not thrive.

3. On the borders of rivers or streams generally.

4. On mountain-slopes, hill-tops, or other elevations.

Forests should be maintained on mountain-slopes, because,—

1. The rainfall is greatest on such slopes.

2. Because the rivers are born in the mountains, and, when the forests are removed, the waters drain so rapidly from the surfaces of the slopes that dangerous floods occur, and much of the soil is rapidly carried away.

3. Because the presence of the forest prevents the occurrence of disastrous droughts.

4. Because the presence of the forest prevents

sudden changes in the temperature of the air, and thus tends to increase the number and severity of hail-storms.

5. Because the presence of the forest tends to prevent the occurrence of early frosts in the neighboring agricultural districts.

6. Because the presence of the forest insures a greater uniformity in the relative quantity of moisture in the air at different seasons of the year.

All life, whether animal or plant, has its beginnings in a minute germ-cell containing a nucleus surrounded by a transparent substance called protoplasm.

Although cases exist where plants appear without the apparent sowing of seed, yet, in all such cases, seeds or germs must have been present.

The conditions necessary for plant-growth, named in the order of their importance, are:

1. The germ or seed.

2. The sunshine and the heatshine.

3. The nourishment, or the food the plant requires for its growth.

4. The cradle, or the soil in which the plant is born.

When a particular species of plant life is to be maintained, the character of the soil is of the greatest importance. But if the other conditions of heat, light, and nourishment exist, almost any soil will be found that will be the best fitted for some few of the great variety of plants.

The germ or seed is in all cases derived from a plant similar to that which is produced when such seed grows and bears fruit.

The soil forms the plant's cradle; in it the plant spreads its roots, and obtains the water and mineral ingredients required for growth.

The moisture and carbonic acid taken from the air by a plant during its active growth form the principal part of the plant's structure; the various mineral matters taken from the soil form but a comparatively small part of such structure.

During active growth in the presence of sunshine, plants take in or absorb carbonic acid from the air. Under the influence of sunlight, this carbonic acid, together with its associated water, is eventually decomposed, the carbon and hydrogen being retained, and the oxygen thrown off into the air.

The mineral matters in the soil must exist in such conditions as will permit of ready assimilation.

Every section of country possesses a nationality

in its plant growth, or produces a particular variety of plants called its flora.

The differences in the distribution of light, heat, and moisture in different parts of the earth cause corresponding differences in the flora of such parts.

The flora of the equatorial regions consists of such plants as are best fitted to exist under the conditions of abundant heat, light, and moisture of these regions.

In passing from the equator to the poles the differences in the distribution of heat and moisture cause corresponding differences in the variety and luxuriance of plant life.

In passing from the base to the summit of a high tropical mountain similar differences in the variety and luxuriance of plant life are noticed, as in going from the equator to the poles.

Seed-time and harvest seldom fail in nature, because the germs of vegetable life are generously scattered in all regions of the earth.

The agencies provided by nature for widely scattering the seeds of plants are various. Some seeds are provided with delicate hair-like wings, which permit the wind readily to carry them great distances from the plants which produced them. Others are provided with hooks or bristles, which catch in the fur of animals or in the plumage of birds, and in this manner are often carried to distant regions.

Some seeds, which are swallowed whole by birds or other animals, often pass out uninjured by the process of digestion at localities far distant from where they were produced.

Civilized man either purposely or accidentally carries seeds from one locality to another.

It sometimes happens that plants introduced into a particular section of country from a distant land find the new soil and climate so favorable to growth as to completely drive out and exterminate domestic species.

The germs or seeds of plants often exhibit a remarkable tenacity of life under certain circumstances. Grains of corn or wheat taken from Egyptian mummies have grown and borne fruit, notwithstanding their centuries of rest.

In a densely-wooded section of country the ground is often so thickly covered by trees as to exclude all other forms of vegetable life. When, however, the removal of a few trees lets in the sunlight and heat, the seeds, which were possibly slumbering in the ground for centuries, at once spring into active life.

When an artesian well is successfully dug in the Sahara Desert, the appearance of the water is almost invariably followed by the appearance of a flora that often contains species peculiar to such districts.

Wherever the virgin soil of the prairies is up-turned, and thus exposed to the air, as by the wheels of the settlers' wagons or other causes, new species of plants appear.

In the North Temperate Zone the burning of pine forests is almost invariably followed by the appearance of scrub-oak.

This wide distribution of plant germs, together with their wonderful vitality, insure the natural growth of a vegetable covering in all regions of the earth where suitable conditions of soil, light, heat, and moisture exist.

The character of the vegetation in any district depends more on peculiarities in the distribution of light, heat, and moisture in such districts than on the character of the soil.

The peculiarities in the distribution of the rain-fall in any country determine to a great extent the character of the flora of such country.

When the rainfall in any region is entirely ab-sent, no matter what the character of the soil may be, or what the amount of light and heat such soil receives, vegetation will be entirely absent, and the region will become a desert.

Where rain falls during one part of the year, and the rest of the year is dry, steppe regions occur. Such regions are covered by vegetation during the wet season, but resemble deserts during the dry season.

Meadows and prairies occur where the rainfall is well distributed throughout the year, and the quantity is not very great.

Forests occur where there is an abundant rainfall well distributed throughout the year.

Forests cannot exist in any part of the North Temperate Zone where the rainfall is absent for a considerable length of time, because the trees would die during the dry season, and there would be no germs for a new crop of trees to start growing from on the appearance of the rainy season.

In certain parts of the tropics forests may exist despite long periods of drought, because in such regions the growth of the trees is either considerably retarded, or the trees obtain their liquid nourishment from copious dews or directly from the vapor of the air.

A certain depth and character of soil are necessary for the growth of trees. Such a soil was formed by the gradual disintegration of hard rocks, and by the growth and subsequent decay of thousands of generations of plants.

Forests are generally found on the slopes of mountains, where the rainfall is considerable and well distributed throughout the year. The mountains are, therefore, the natural home of the forest.

Forests are found especially on that coast of an island or continent which is exposed to the prevalent wind, because there the rainfall is considerable and no extended time occurs when the rain is absent.

Soil was originally formed by the gradual disintegration of the crystalline rocks that were produced by the cooling of the earth's crust.

Disintegration of rocks is effected by various causes, mainly, however, by the action, in some way or another, of water.

Sometimes the soil is found resting on the surface of the rock from which it was derived by disintegration. In such cases its general character can be directly traced to the composition of the underlying rocks by the gradual change which can be observed from the loose, porous soil on top, to the hard, untouched, virgin rock below.

Soils may be divided into gravelly, sandy, clayey, calcareous, and peaty.

The agencies by which the hard crystalline rocks may be broken up or disintegrated to form soil are,—

1. The expansion produced during the sprouting or growing of vegetation.

2. The alternate contractions and expansions that attend the freezing or thawing of the water that sinks into the rocks.

3. The cutting or eroding power of running-water charged with suspended mineral matters.

4. The eroding or cutting power of glaciers.

5. The solvent power of water containing such gases as oxygen or carbonic acid.

A plant, during its vigorous growth, by the expansion of its roots, may break or rend a rock, and thus aid in its disintegration.

The alternate expansions and contractions that attend the thawing or freezing of the water which sinks into a rock gradually break the rock into fragments, and thus aids in the formation of soil.

During the gradual movements of glaciers down the mountain valleys, the fragments of hard rocks lodged in the ice cut or grind the rocks which form the sides of the valleys through which the glaciers move, and thus aid in the formation of soil.

When water contains dissolved in it oxygen or carbonic acid gas, it may gradually dissolve some of the less insoluble ingredients of the hardest rocks, and thus cause them to become permeable.

Clayey soils are derived from the disintegration of feldspathic rocks.

Calcareous soils are derived from the disintegration of limestones.

Some soils possess the valuable property of absorbing moisture directly from the vapor in the air. Soils containing a large quantity of vegetable humus possess this property in a more marked degree than any others. Clayey soils also possess it to a marked extent.

The ability of soils to absorb the sun's heat will vary with their color. Dark-colored soils absorb the heat much better than light-colored soils.

The plants that are found growing naturally in any locality are those which are best fitted to grow in such locality. They will continue to grow naturally only so long as these favorable conditions are maintained.

Forests require for their continued existence a certain character of soil, so that, although even all the climatic conditions requisite for their growth exist, they cannot appear until such soil is provided.

Like other forms of creation, the forest is forced to maintain a continual struggle for existence.

Its enemies may be divided into two classes,— namely, animate and inanimate.

The principal animate enemies of the forest are plants, animals, and man.

The principal inanimate enemies of the forest are fire, winds, floods, and avalanches.

The destruction of the forest by fire is generally complete. Though in some cases a small fire may, by destroying the less hardy forms of plant life, increase the growth of certain trees, such, for example, as the pitch-pine, yet, in general, extensive forest fires generally so completely remove the forests, that it is often impossible to re-establish them.

Severe forest fires generally occur during the dry season of the year. The rain which subsequently falls finds the ground unprotected by any vegetable covering, and, rapidly draining off the surface, carries away much soil.

The principal causes of forest fires are the camp-fire, the burning of brush, the locomotive spark, the lightning-bolt, and at times, perhaps, the heating power of the sun's rays concentrated by lenticular, resinous, or gummy nodules.

When its velocity is great, the wind sometimes sweeps away the trees from extended areas. This action of the wind is limited mainly to the edges of the forest or to openings made in them by any cause.

By overflowing their banks, rivers sometimes undermine and carry away thousands of acres of forest trees. The trees accumulate in the bed of rivers and form masses called rafts.

An avalanche sweeping down the slope of a mountain often completely removes the forest. A plot of forest land, properly placed, will often check the movements of avalanches.

The animate enemies of the forest often produce their greatest destruction by the aid of the inanimate enemies. Thus, man destroys forests by fire; the beaver, by floods.

Various parasitic plants may grow on, and thus cause the death of, even the most vigorous trees.

Some forms of fungus-growth cause considerable damage to the trees on which they grow.

The animal enemies of the forest vary in size from minute insects to animals of large size.

The ravages of the animal kingdom are most marked on the borders of the forest. In the deeper recesses, the vegetable kingdom holds almost undisputed sway, excluding the animal forms by the density of its growth.

Domestic animals, when allowed to range freely through the forest, may cause considerable damage, by destroying the foliage, or by gnawing the bark of trees.

Among wild animals, rodents are the most destructive by gnawing the bark, and often by completely girdling the trees.

Rabbits, mice, and beavers are among the rodents that cause the greatest damage to the forests.

Beavers destroy forests not only by actually cutting down trees, but also by building dams which cause the overflowing of the adjacent country, which thus destroys the timber growing thereon.

Goats and other animals that live largely on the bark of trees often work great destruction to the forests.

Insects cause damage to the forests, either by feeding on the parts of the tree necessary for reproduction or growth, or by making galleries or tunnellings through the wood. Many insects, while in the larva state, cause great damage to trees by boring or eating the wood.

Various caterpillars often cause so great a destruction to the pine-trees as to completely destroy extensive pine forests.

The greatest enemy of the forest is man. As lord of the forest he is entitled to its products, and if he exercises judgment he can safely harvest his forest crops.

The removal of the forests from any considerable section of country is almost invariably attended by some or all of the following results,—namely:

1. An increase in the frequency and severity of the inundations of the rivers flowing in or through such districts.

2. An increase in the number and severity of droughts in such districts.

3. A rapid loss of the soil of the area from which the trees have been removed.

4. A marked disturbance in the lower courses of the rivers which rise in, or flow through, such districts.

5. An increase in the number and severity of hail-storms.

When the forests are removed from any area,

the rain which falls on such area, instead of slowly draining into the river channel during a comparatively long time, drains rapidly into it and causes disastrous floods. The reservoirs of the springs in such districts thus failing to receive their proper supply of water, are apt to dry up shortly after the beginning of the drought.

The rapid drainage of the areas from which the forests were removed causes a loss of its surface-soil.

The soil thus lost to the highlands is deposited in the lower courses of the rivers, in the shape of mud-flats, or sand-bars, which injuriously affect navigation.

When soil, rich in vegetable humus, deposited on the lowlands near the mouths of rivers, is exposed to the sun's heat, is apt to cause miasmatic or other diseases.

The ground left bare by the destruction of the forest permits it to both take in and part with its heat rapidly, and thus to permit the air over it to rapidly grow hot in summer and cold in winter.

Forests should be maintained in some parts of all regions where trees can grow. The best places for such purposes are to be found on the slopes of mountains.

From every water surface of the earth vapor is almost constantly passing into the atmosphere. This vapor diffuses through the air over such water surfaces, and is carried by the winds to different regions of the earth.

The heat which turns water into vapor disappears, or becomes what is commonly called latent heat. When such vapor is sufficiently chilled and falls as rain, snow, or other form of precipitation, the latent heat becomes sensible and warms the surrounding air.

The rapidity with which water evaporates or passes into the air as vapor varies with the following circumstances:

1. The extent of the surface exposed.
2. The temperature of the air.
3. The quantity of vapor already in the air.
4. The pressure.

The vapor which passes into the air exerts a considerable influence in moderating the extreme temperatures that would otherwise exist in the equatorial and polar regions of the earth, in the following ways, viz.:

1. By effecting an interchange between the excessive heat of the equatorial regions and the excessive cold of the polar regions.

2. By acting as a screen which both prevents the earth's surface from being too rapidly heated on exposure to the sun's rays, or too rapidly cooled when deprived of such rays.

Since air can hold more vapor when hot than when cold, if the temperature of a mass of warm moist air is sufficiently cooled, the moisture it can no longer hold as vapor appears as rain or as some other form of precipitation.

The lowering of temperature required to produce rain is obtained:

1. By warm moist air blowing along the earth's surface towards colder regions.

2. By warm moist air rising directly from the earth's surface into the higher and colder regions of the atmosphere.

Rain is generally caused by warm moist air blowing towards the polar regions of the earth. Cold dry air blowing towards the equatorial regions has its capacity for moisture increased, and tends rather to cause droughts than rain.

In tropical regions, a wind that has crossed an ocean, and has thereby become saturated with moisture, may bring rain on reaching the coast of a continent or island, in no matter from what direction it comes.

More rain falls in the equatorial regions than elsewhere; more falls in the temperate than in the polar regions. More rain falls on the coasts of continents than in the interior.

Where the temperature is sufficiently high, as in the equatorial regions, rain may be caused by the chilling produced by ascending currents.

Mountains cause a heavy rainfall on account of the air being chilled when forced to ascend their cold slopes.

Nearly all the great rivers of the world rise in mountainous districts.

The rain that falls on the earth either runs directly off the surface or sinks slowly into the ground. The part that runs directly off the surface collects in streams that discharge directly into the rivers. The part that sinks into the ground collects in underground basins, from which it slowly emerges as springs.

When the mountains are covered by forests, the rain which falls on their slopes, for the greater part, drains slowly into the ground. When the mountains are denuded of their forests most of the rain drains rapidly off the surface. The destruction of the forests on mountain-slopes is, therefore, apt to cause floods.

The running of the water from a higher to a lower level is called drainage.

There are two kinds of drainage:

1. Surface drainage, where the water runs directly off the surface into the rivers.

2. Underground drainage, where it first sinks into the ground and afterwards discharges as springs into the rivers.

Underground drainage takes place slowly. Surface drainage takes place rapidly.

The direction in which rivers flow depends on the direction in which the land slopes.

The main stream, with all its tributaries and branches, is called the river system. The land which drains into a river system is called the river basin. The size of the river depends on the ratio between the quantity of the rainfall and the size of the river's basin.

When a river basin is covered with a loose, porous soil, such as it will have when covered with almost any form of vegetation, the rain sinks slowly into the earth and the river seldom overflows its banks.

When the area is such that most of the water runs directly off the surface, as will generally be the case when deprived of its vegetable covering, the rivers receiving such drainage are apt to over-flow their banks during the wet season.

Any disturbance in the natural drainage of a country may cause damage from the too rapid drainage of its surface:

1. By insuring too much water in its rivers during inundations.

2. By insuring too little water in its rivers during drought.

The preservation of forests on mountain-slopes, where the drainage is more rapid than elsewhere, insures a proper drainage of its surface, and consequently the proper flow of its rivers.

The condition of the air of a country as regards its heat or cold moisture or dryness, healthfulness or unhealthfulness, is called its climate.

The atmosphere receives its heat from the sun:

1. Directly, by absorption.

2. Indirectly, from the heated earth.

The atmosphere receives its heat indirectly from the heated earth:

(a) By contact with the heated earth.

(b) By radiation from the heated earth.

(c) By reflection from the heated earth.

All parts of the earth in the same latitude do not possess the same temperature, because the

surface is higher in some places than it is in
others, in some places is covered with vegetation
and in others is bare, or is exposed to cold or
currents of wind or water in some places and to
warm currents in others.

Differences in the elevation of the land cause
differences in the temperature of the air. An
elevation of three hundred and fifty feet will cause
the same lowering of temperature as a difference
of one degree of latitude,—viz., of 1° Fah.

The same changes of temperature are observed
in passing from the base to the summit of a high
tropical mountain as in passing along the earth's
surface from the equator to the poles.

Portions of the earth's surface covered by water
heat or cool slowly; consequently, the air over
such portions does not change its temperature
rapidly, or, in other words, such portions of the
earth possess an equable climate. Portions of
the earth covered by land heat and cool rapidly;
consequently, the air over such portions changes
its temperature rapidly, or, in other words, such
portions possess a variable climate.

A surface covered with vegetation—such, for
example, as a forest—does not change its temperature as rapidly as it would if it were bare. Forests,

therefore, tend to prevent sudden changes in the climate.

A bare, uncovered area, such as a desert, is subject to sudden changes in its climate.

The climate produced by an extended land area is called a continental climate; that produced by an extended water area, an oceanic climate. A continental climate is characterized by great and sudden changes of temperature; an oceanic climate, by a comparatively uniform temperature. The forests tend to produce a climate characterized by a comparatively uniform temperature. In this respect, therefore, the forest climate is like the oceanic climate.

The sun does not heat an area covered by forests either as intensely or as rapidly as a bare area, because :

1. The heat is spread over the greatly-extended surfaces formed by the trees of the forest and its underbrush.

2. The vegetable covering acts as a screen to protect the ground from the direct action of the sun's rays.

3. The air over the forest is moister than that over the fields, and this moist air acts either as a screen to protect it from the heat of the sun,

or to prevent the loss of its own heat by radiation.

Therefore, an area of ground covered with forests is subjected to smaller changes of temperature
than a bare, uncovered area.

The climate of the forest is more equable than
that of the open fields, because the forest takes
in and parts with its heat more slowly than the
fields.

A layer of snow tends to preserve the temperature of the ground on which it falls. If snow
falls on unfrozen ground, the ground will probably
remain unfrozen throughout the year until the
snow melts; and, when the melting occurs, the
water will drain slowly into the earth. If, however, the snow falls after the ground is frozen, the
ground will probably remain frozen until the snow
melts, when the water will drain rapidly off the
surface.

The presence of the forest tends to keep the
ground unfrozen until protected by a layer of
snow, and in this way, when the snow melts, the
water sinks quietly into the ground, and disastrous
floods are thus avoided.

The forest, by keeping the air over it moister than
that over the fields, increases the ability of the air

to take in heat, either directly from the sun, or indirectly from the heated earth.

Forests prevent sudden changes of temperature throughout the year. In early autumn they decrease the frequency of destructive frosts by preventing the temperature of the air from rapidly falling.

The presence of forests over extended areas prevents the occurrence of sudden changes of temperature.

1. By permitting such areas to more thoroughly absorb the sun's heat, on account of the greater surfaces they possess.

2. By keeping the air over the forests moister than over the open fields, thus enabling it more readily to absorb the sun's heat.

3. By acting as a screen to the lands lying to the leeward of cold winter winds.

4. By preventing the frosts from penetrating great distances into the ground, and, therefore, increasing the chance of winter snows falling on unfrozen ground.

The atmosphere is composed of a mixture of about seventy-seven per cent. by weight of nitrogen and twenty-three per cent. of oxygen. It also contains a nearly constant quantity of carbonic acid gas and a variable quantity of the vapor of water.

The oxygen of the air is necessary for animal life. The carbonic acid gas is necessary for plant life. The moisture is necessary for both animal and plant life.

Animals take in oxygen and give out carbonic acid gas. During growth, when exposed to sunshine, plants take in carbonic acid gas and give out oxygen.

The presence of both animal and plant life, therefore, is necessary to keep the composition of the atmosphere the same.

In the geological past the earth's atmosphere was vaster than at present. It contained more oxygen and more carbonic acid than it does now. Much of the oxygen, which then existed in a free state in the air, is now combined with various materials that form the earth's crust.

The excess of carbonic acid which existed in the earth's atmosphere during the geological past was removed from it mainly by the plants of the carboniferous period, and now exists in the earth as beds of coal.

In order to avoid any disturbance in the balance between plant and animal life of the earth, forests, which represent the largest forms of plant life, should be preserved.

Hail occurs when considerable differences of temperature exist between neighboring masses of very moist air.

The destruction of the forest, by readily permitting such differences of temperature to occur, tends to increase the number and severity of hail-storms.

A hail-storm is generally preceded by the appearance of several layers of dark grayish clouds, and a violent movement is often seen to occur between them, that is probably of a whirling character.

Hail-storms are almost invariably attended by marked disturbances in the electrical equilibrium of the atmosphere.

A hailstone is formed of alternate layers of ice and snow. Various explanations have been offered to account for the peculiar structure. Volta ascribed it to the alternate attractions and repulsions occurring between neighboring snow- and rain-clouds, when charged with opposite kinds of electricity.

In France, where Volta's theories were formerly received, lightning-rods were fruitlessly erected on the fields in order to protect them from the ravages of the hail.

The peculiar structure of the hailstone has also been ascribed to a whirling motion of the air between snow- and rain-clouds around a horizontal axis, whereby particles of snow are carried alternately into the rain- and snow-clouds, and thus receive their alternate coatings.

Another theory accounts for the alternate coatings of ice and snow by the exposure of moisture to the different temperatures occurring in denser and rarer portions of space around which the wind is whirling.

By reforestation is meant the replanting of trees in any locality from which they have been removed either accidentally or purposely.

Provided the removal of the forest has not been attended by too great a loss of soil, the same kind of trees may be successfully replanted.

There are two methods by means of which reforestation may be effected.

1. Sowing or Seeding.

2. Tree Planting.

Seeding can be profitably followed in the temperate latitudes where the growth of the tree is comparatively certain.   In higher latitudes the planting of the tree is, perhaps, preferable, since the germination and continued growth of seeds are by no means certain.

Where the destruction of the forest has been caused by an avalanche, the removal of the soil, in some cases, is so complete that trees cannot be successfully replanted.

In some of the western parts of the United States it is now recognized, from actual experience, that when trees are planted in plots around the farm-lands, the protection thus afforded the rest of the farm-land, against the winds, is of greater money-value than the rent of the ground occupied by such trees.

The following locations are especially adapted to tree planting :

1. Wet lands.

2. Lands covered with a thin or poor soil.

3. Along the margins of rivers where the land is not required for roads or other public purposes.

4. On the side of mountain-slopes where the soil is of the proper character, or subject to destructive avalanches.

An exact balance must be preserved between the five great geographical forms,—namely, the land, the water, the air, the plants and animals, in order that the complex organization of nature may be properly maintained in operation.

The energy which is the cause of nearly all natural phenomena of the earth is received directly from the sun.

One part of the sun's heat stirs the air and water in vast movements between the equator and the poles, and thus effects an interchange between the too great heat of the equatorial regions and the too feeble heat of the polar regions. Another part of the solar energy is directly expended in producing one or another of the myriad forms of plant or animal life.

If the land, the water, or the air receives more than its share of solar energy, a disturbance in the balance of nature is effected, which produces marked effects in the life of the earth.

The total water area of the earth bears a proportion to its total land area very nearly as 25 is to 9, or as $2\frac{7}{9}$ is to 1.

The earth receives its greatest heat from the sun at those parts of its surface where it has its greatest water areas, only a comparatively small part of the land being found in the equatorial regions. There are produced, however, vast currents in the atmosphere and in the ocean, which effect an interchange between the heat of the equator and the cold of the poles.

Any change in the distribution of the land and water areas of the earth, either as regards their relative amount, or as regards their distribution, would seriously affect the life of the earth.

If the greatest proportion of land existed at the equator, such changes would be produced in the earth's climate as to sweep its present life out of existence.

Any change in the elevation of the present land areas would produce a marked change in the earth's climate. It was probably an increase in the elevation of the polar lands that caused the severe cold of the glacial epoch, when so much of the northern continents were covered with ice.

In order to preserve the present relative proportions of oxygen and carbonic acid in the atmosphere, the present animal and plant life of the earth must be preserved.

Animals are absolutely dependent on plants for their existence. The plants can take their food directly from the air and the soil. Animals require their food to be prepared for them by the plants on which they live. The death of the plant life of the earth would, therefore, be followed by the death of all its animals.

Many animals multiply so rapidly that, unless they were removed from the earth by furnishing food for other animals, a marked disturbance would be effected in the balance of nature.

APPENDIX.

LISTS OF TREES SUITABLE FOR REPLANTING IN DIFFERENT PORTIONS OF THE UNITED STATES.

In order to extend the scope of the OUTLINES OF FORESTRY, and to render it of greater practical value, the following circular letter was sent to different well-known authorities in forestry, inquiring as to lists of trees suitable for replanting in different sections of the United States.

Circular Letter.

PHILADELPHIA, January 23, 1892.
1809 Spring Garden Street.

PROF. ....

....

....

DEAR SIR,—I am about publishing a little work on Forestry, and am desirous of obtaining a list of trees suitable for planting for reforestation in such parts of.................................... and the adjoining States as have been denuded of forests, or are capable of sustaining forest trees.

In the event of your being able to spare the time necessary

to send me a list of such trees, I would, of course, make full acknowledgment in the book of your contribution.

If your time is too fully occupied to send me the information requested, can you inform me where I can obtain the same?

Asking your pardon for thus trespassing on your valuable time, I am,

Very respectfully, yours,

EDWIN J. HOUSTON.

The letters received in reply to the foregoing, together with the lists of trees suitable for purposes of reforestation, are hereunto appended.

From Thomas Meehan, Editor of Meehan's Monthly, German-town, Philadelphia.

MEEHAN'S MONTHLY,
GERMANTOWN, PHILADA., January 30, 1892.

PROF. EDWIN J. HOUSTON,

    1809 Spring Garden Street, Philadelphia.

DEAR SIR,—I have marked in the catalogue sent to-day the names of such trees as are most desirable for planting on the Northeastern Seaboard of the United States. A few are rare, but will soon become common; others are not likely to become common for some years. In some cases the trees marked would have but limited usefulness, but all are of value in some respect or another.

When getting into the States along the seaboard of the Virginia line, many of those named would be ineligible.

Very truly yours,

THOMAS MEEHAN.

DECIDUOUS TREES.

MAPLES.

Acer campestre, or European Cork Maple.

" *palatanoides,* or Norway Maple.

" *pseudo-platanus,* or European Sycamore Maple.

HORSE-CHESTNUT.

Æsculus glabra, or American Horse-Chestnut.

" *hippocastanum,* or European Horse-Chestnut.

AILANTUS.

Ailantus glandulosa.

BIRCH.

Betula alba, or European White Birch.

HICKORY.

Carya alba, or Shell-bark Hickory.

" *amara,* or Bitternut Hickory.

" *microcarpa,* or Small-Fruited Hickory.

" *sulcata,* or Large-Fruited Hickory.

" *tomentosa,* or White Hickory.

SWEET CHESTNUT.

Castanea Americana, or American Chestnut.

HOLLY.

Ilex opaca, or American Holly.

JUNIPER, CEDARS.

Juniperus Virginiana, or Red Cedar.

FIR.

Picea balsamea, or Balsam Fir.

PINE.

Pinus Austriaca, or Australian Pine.

" *Banksiana.*

" *densiflora,* or Japan Pine.

" *Laricio,* or Corsican Pine.

Pinus Massoniana.
" *pungens.*
" *resinosa,* or Red Pine.
" *rigida,* or Pitch Pine.
" *mitis,* or Yellow Pine.
" *strobus,* or White Pine.

JAPAN CYPRESS.
Retinispora obtusa.

ARBOR-VITÆ.
Thuja occidentalis, or American Arbor Vitæ.

LOCUST, ACACIA.
Robinia pseudacacia, or Yellow Locust.

MAIDEN-HAIR TREE.   GINGKO.
Salisbùria adiantifolia.

WILLOW.
Salix alba, or White Willow.
" *Babylonica,* or Weeping Willow.
" *japonica,* or Japan Willow.
" *pentandra,* or Laurel-leaved Willow.
" *Russelliana.*
" *vitellina,* or Golden-Bark Willow.

ELMS.
Ulmus Americana, American Elm.
" *campestris,* or English Elm.
" *fulva,* or Slippery Elm.
" *racemosa,* or American Cork Elm.

EVERGREENS.

SPRUCE.
Abies alba, or White Spruce.
" *Canadensis,* or Hemlock.

Abies Douglasii, or Douglas Spruce.

" *excelsa,* or Norway Spruce.

" *pungens,* or Colorado Blue Spruce.

LARCH.

Larix Europœa, or European Larch.

SWEET GUM.

Liquidambar styraciflua.

SOPHORA.

Sophora Japonica.

DECIDUOUS CYPRESS.

Taxodium distichum.

LINDEN.

Tilia Americana, or American Linden.

" *Europœa,* or European Linden.

POPLAR.

Populus alba, or Silver Poplar.

" *balsamifera,* or Balsam Poplar.

" *fastigiata,* or Lombardy Poplar.

" *grandidentata.*

" *monilifera,* or Carolina Poplar.

OAK.

Quercus alba, or White Oak.

" *bicolor,* or Swamp White Oak.

" *cerris,* or Turkey Oak.

" *coccinea,* or Scarlet Oak.

" *dentata* (Daimio), or Japan Oak.

" *falcata,* or Spanish Oak.

" *imbricaria,* or Laurel Oak.

" *macrocarpa,* or Mossy Cup.

" *nigra,* or Black Jack Oak.

" *obtusiloba,* or Post Oak.

Quercus palustriis, or Pin Oak.

    "   *phellos,* or Willow Oak.

    "   *prinus,* or Rock Chestnut Oak.

    "   *robur,* or English Oak.

    "   *rubra,* or Red Oak.

    "   *tinctoria,* or Black Oak.

MULBERRY.

Morus alba, or White Mulberry.

    "   *rubra,* or American Red Mulberry.

BOX ELDER.

Negundo fraxinæfolium, or Ash-leaved Maple.

SOUR GUM.

Nyssa multiflora.

IRONWOOD.

Ostrya Virginica.

EMPRESS-TREE.

Paulownia imperialis.

BUCKEYE.

Pavia flava, or Yellow Buckeye.

CHINESE CORK-TREE.

Phellodendron amurense.

PLANERA.

Planera cuspidata.

BUTTONWOOD, PLANE.

Plantanus occidentalis, or American Plane.

    "   *orientalis,* or Oriental Plane.

HONEY LOCUST.

Gleditschia triacanthos, or Honey Locust.

KENTUCKY COFFEE.

Gymnocladus Canadensis.

HOVENIA.
Hovenia dulcis.

IDESIA.
Idesia polycarpa.

WALNUT.
Juglans cinerea, or Butternut.
    "   *regia,* or English Walnut, or Madeira Nut,—south
        of Philadelphia.

TULIP-TREE.
Liriodendron tulipifera, or Common Tulip-Tree, or Tulip
        Poplar.

OSAGE ORANGE.
Maclura aurantiaca.

MAGNOLIA.
Magnolia acuminata, or Cucumber-Tree.
    "   *macrophylla.*
    "   *tripetala,* or Umbrella-Tree.

CATALPA.
Catalpa bignonioides, or Catalpa.
    "   *speciosa,* or Western Catalpa.

NETTLE-TREE.
Celtis occidentalis, or Nettle-Tree.

CHERRY.
Cerasus avium alba plena, Double-flowering Cherry.
    "   *Pennsylvanica,* or Wild Red-Cherry.
    "   *ranunculæflora.*

DOGWOOD.
Cornus Florida, or White or Large-flowering Dogwood.

PERSIMMON.
Diospyros Virginiana, or American Persimmon.

BEECH.

Fagus Americana, or American Beech.
" *sylvatica,* or European Beech.

ASH.

Fraxinus Americana, or White Ash.
" *excelsior,* or European Ash.
" *quadrangulata,* or Blue Ash.
" *sambucifolia,* or Black Ash.
" *viridis,* or Green Ash.

KATSURA.

Cercidiphyllum japonicum.

From B. S. Hoxie, Secretary of the Wisconsin State Horticultural Society.

WISCONSIN STATE HORTICULTURAL SOCIETY,
EVANSVILLE, WISCONSIN, January 26, 1892.

EDWIN J. HOUSTON, Philadelphia, Pa.

DEAR SIR,—Yours of the 28th at hand. I mailed to your address our last volume, No. 2, I think. The list of trees found on page 9, which we recommend for general planting, will be found applicable to nearly all parts of our State.

The pine regions of Wisconsin are the parts that are now being deforested, and no special effort is being made to reforest these areas. Oak, ash, maple, birch, elm, pine, and spruce will grow on most of this land.

We have several townships bordering on Lake Superior in Bayfield County, which some ten years ago were set apart as a State Park. Last winter a bill was introduced to bring this reservation into the market,—*i.e.,* to sell the timber, but it failed to pass, and if my pen and the press can prevent such a bill, it will never pass. If I can be of any further assistance to you, please write me.

I shall be glad to see your book when published.

Respectfully,

B. S. HOXIE.

TREES AND SHRUBS RECOMMENDED.

Trees and shrubs recommended in the "Annual Report of the Wisconsin State Horticultural Society." *

EVERGREENS.

For general planting in the order named:

> White Pine.
>
> Norway Spruce.
>
> White Spruce.
>
> Arbor Vitæ.
>
> Balsam Fir.
>
> Austrian Pine.
>
> Scotch Pine.

For ornamental planting in the order named:

> Hemlock.
>
> Red Cedar.
>
> Siberian Arbor Vitæ.
>
> Dwarf Pine.
>
> Red or Norway Pine.

---

* Annual Report of the Wisconsin State Horticultural Society, embracing papers read and discussions thereon at the semi-annual meeting held in Black River Falls, June 26, 27, 1890; also at Madison, June 2–6, 1891. Vol. xxi. p. 9.

DECIDUOUS TREES.

For Timber.

White Ash.
Black Walnut.
Hickory.
Black Cherry.
Butternut.
White Oak.
European Larch.
American Larch.

Street Shade-Trees.

White Elm.
Hard Maple.
Basswood or Linden.
Ashleaf Maple (*Acer negundo.*)
Norway Maple.
Hackberry.

For Lawn Planting.

Weeping Cut-leaved Birch.
American Mountain Ash.
Green Ash.
Horse-Chestnut.
European Mountain Ash.
Wisconsin Weeping Willow.
Oak-leaved Mountain Ash.
White Birch.
Weeping Golden-barked Ash.
Weeping Mountain Ash.
Weeping Poplar.

From Charles Mohr, Agent for the Forestry Division of the United States Department of Agriculture, Mobile, Alabama.

U. S. DEPARTMENT OF AGRICULTURE,
MOBILE, ALABAMA, January 31, 1892.

PROF. EDWIN J. HOUSTON, Philadelphia, Pa.

DEAR SIR,—Your letter of the 23d has been received. According to your request, I send you enclosed a list of timber trees, which might be regarded as adapted for the reforestation of denuded areas in the Gulf States east of the Mississippi River.

In the selection of the trees, I had to be guided solely by my observations made in the different sections of the regions named, and had to confine myself entirely to native species, no information being on hand in regard to trees from other sections of the United States, or exotics.

To shorten matters, I refer you for information about the habits of the species named in the list, to the preliminary of important forest trees in the United States, in Mr. Fernow's "Report to the Commissioners of Agriculture" (Forestry Division) for the year 1886, where, also, notes on the economic uses of each will be found. The numbers in my list refer to the same species mentioned in the above report.

I remain truly yours,

CHARLES MOHR.

White Cedar, *Chamæcyparis sphæroidea.*
Red Cedar, *Juniperus Virginiana.*
Bald Cypress, *Taxodium distichum.*
Long-leaved Pine, *Pinus palustris.*
Loblolly Pine, *Pinus tæda.*

Cuban Pine, *Pinus Cubensis.*

Short-leaved Pine, *Pinus mitis.*

White Oak, *Quercus alba.*

Cow Oak,   "   *Michauxii.*   Rich, alluvial soil.

Chestnut Oak, "   *prinus.*

Live Oak,   "   *virens.*   Lower districts.

Red Oak,   "   *rubra.*

Black Oak,   "   *tinctoria.*   Gravelly uplands.

Spanish Oak, "   *falcata.*   Throughout on lighter soils.

Water Oak,   "   *aquatica.*   Of value for fuel only.

Willow Oak,   "   *Phellos.*   Rapid growth on wet or dry
                        light soil.   Timber more valuable.

Beech, *Fagus ferruginea.*

Chestnut, *Castanea vulgaris*, var. American.   In dry, some-
     what silicious soils throughout.

Shell-bark Hickory, *Carya alba.*   Upper and central districts.

Mocker Nut, *Carya tomentosa.*

Pecan, *Hickoria Pecan (Carya olivæformis)*.   Valuable for its
     fruit.

Wild Cherry, *Prunus serotina.*

Sweet Gum, *Liquidambar Styraciflua.*

Black Locust.

Honey Locust.

Red Mulberry, *Morus rubra.*   Shade-loving.

Magnolia, *Magnolia grandiflora.*

Cucumber Tree, *Magnolia acuminata.*   In well-drained, deep,
     and lighter loamy soil, through the State.   Shade-loving.

Tulip-Tree, *Liriodendron tulipifera.*

Osage Orange, *Maclura aurantiaca.*

White Ash, *Fraxinus Americana.*   Upper district.

Green Ash, *Fraxinus viridis.*

Red Maple, *Acer rubrum.*

Silver Maple, " *dasycarpum.*

White Elm, *Ulmus Americana.* Upper district.

Slippery Elm, " *fulva.*

Water Elm, " *alata.* Alluvial, wet soil.

American Linden, *Tilia Americana.* Central to upper district.

Sycamore, *Plantanus occidentalis.*

Cottonwood, *Populus monilifera.*

From Robert W. Furnas, Secretary of the Nebraska State Board of Agriculture.

NEBRASKA STATE BOARD OF AGRICULTURE,
;BROWNVILLE, NEBRASKA, January 27, 1892.

EDWIN J. HOUSTON, ESQ.,
1809 Spring Garden Street, Philadelphia, Pa.

DEAR SIR,—Reply to yours, First, twenty-third. Of the more valuable hard-wood varieties of timber used for forestry purposes on our prairies, or naturally timberless region, we find the best,—black walnut, white ash, black and honey locust, black cherry, Kentucky coffee-tree, hard maple, burr and white oaks.

Of the soft woods,—soft maple, box elder, cottonwood, and the catalpas,—speciosa, and Tea's hybrid.

Evergreens: red cedar, Scotch, Australian, and white pines. Both American and European larch do well.

If I can serve the forestry cause in any way, command me.

Truly,
ROBERT W. FURNAS.

19*

From Professor M. G. Kern, Editor of Coleman's Rural World.

St. Louis, January 26, 1892.

EDWIN J. HOUSTON, ESQ., Philadelphia, Pa.

DEAR SIR,—Your favor of the twenty-third inst. received.

Kindly excuse a few remarks on Western forestry, rather outside of the list of available trees requested. But a part of Missouri is heavily timbered, and in these sections the great object of the population is to get rid of the timber for money or for agricultural purposes. Many portions of the State are cleared to the same extent as the older States, with abundance of forest supplies for present use. Part of the State is prairie, in which the need of timber culture is as great as in all open plains. South of us is Arkansas, a mountainous native forest; west, Kansas, Nebraska, and North Iowa, with her endless prairies. A wide field, indeed, for forest management, reforesting, and timber culture. The soil brings forth, mostly with rapid growth, all the indigenous species of our forests; a list of such could be valuable only in ratio to the value of the timber grown.

Fast-growing soft woods are considered in wooded sections as of little value, scarcely worth planting on a large scale. They are generally planted for shelter, ornament, or quick shade. No line of difference as to hardiness can in reality be drawn, though variety of conditions make certain kinds more suitable to certain localities: we differ in this respect from more northern latitudes.

The question of reforesting denuded sections has thus far attracted but little public attention. Intelligent husbandry steadily advancing, is gradually awaking to the necessity of reforestation and the protection of native forests. The prospective value of economic timber is realized by many, and the growth of young native timber is encouraged by the usual means of protection from cattle and disastrous fires.

Systematic forest culture, for which there is so boundless a field, has thus far been practised to but small extent, though every intelligent citizen knows that the most valuable kinds of timber are being exhausted very fast. How long this spell of popular indifference will last cannot be foretold by the wisest. Millions of acres in the lowlands and on hill-sides capable of producing the most valuable timbers, lie idle as though belonging to some Indian tribe.

Much popular education is surely needed to break the barriers of indifference and popular selfishness underlying all the evil. We need a tow-line attached to the dormant intelligence of the people (as far as American forestry is concerned), to draw it into action. When once aroused from the dream of inexhaustible forest wealth, the West will do its honest share in forest culture, as her resources are almost without limit.

My most sincere wishes for the success of your forthcoming work on Forestry. Command my services at any time when special features of information from this section are desired.

Very respectfully,

M. G. KERN.

TREES PRINCIPALLY GROWN IN FOREST CULTURE IN MISSOURI AND ADJOINING SECTIONS.

Initial Step in Forest Culture.

  Cotton-wood.

  Black Walnut.

  Soft Maple.

Box Elder.

White American Ash.

Sugar Maple (to some extent).

Second Step.

The introduction of Catalpa species. At present the above with Black Walnut and White Ash are the leading trees.

Worthy of extensive culture, though sparsely planted thus far, are:

Tulip-Tree (Yellow Poplar).

The leading White Oak species.

White Oak.

Over-cup Oak.

Burr Oak.

Post Oak.

White Hickory (under certain conditions).

Of Conifers.

Bald Cypress (for alluvial lowlands).

Red Cedar, for limestone uplands and stony slopes.

White Pine, especially suited for sandstone formations found in sections of Southwestern Missouri and Arkansas. Ornamental trees of this species, met everywhere, show remarkable vigor of growth and adaptation to the formation.

Black Cherry (to be recommended for general culture).

European Alder, for moist and loamy soils, of rapid growth, and valuable wood. One of the best of temporary nursery trees for larger plantations.

Soft Woods.

Soft Maple.

Box Elder.

Birch (both nigre and lenta).

Willows.

Poplars.

Plantanus, Elms, Sycamore, for various purposes of shade, shelter, and wind-breaks.

Evergreens.

American White Spruce deserves far more attention than thus far enjoyed. It is the most lasting of all spruces.

Norway Spruce.

Scotch and Austrian Pine are valuable for close plantations for shelter. They are, however, deficient as to longevity, losing their vigor much sooner than the native American species.

European Larch is suited to the northern parts of Missouri, Iowa, and farther north.

UNIVERSITY OF PENNSYLVANIA,
PHILADELPHIA, PA., February 27, 1892.

PROF. EDWIN J. HOUSTON,

     1809 Spring Garden Street, Philadelphia, Pa.

MY DEAR PROFESSOR,—I am in receipt of your favor of the 20th inst., and I take pleasure in sending you a list of trees suitable for planting in the Southwestern States and California.

Betula occidentalis, Hook. Black Birch. For ornamental purposes, grows ten to twenty feet high.

Quercus undulata, Torr. Oak. Four varieties, valuable for timber and for masting, furnishing, as they do, an abundance of sweet, edible acorns.

Salix cordata, Muhl., var. *vestita,* Anders, Diamond Willow, abundant in Yellowstone regions, valuable for furnishing unique canes.

p

Salix purpurea, L. Purple Willow, six to fifteen feet. Best of hedge and osier willows. Correspond with United States Department of Agriculture concerning.

Salix lucida, Muhl. Shining Willow, five to ten feet. One of the most beautiful willows for ornamental planting.

Salix fragilis, L. Brittle Willow, Crack Willow, Bedford Willow, sixty to eighty feet, affords best willow timber, and contains large per cent. of tannin, and more salicin than others.

Salix alba, L., vars. Salix cæruea and Salix Vitellina, Blue Willow, and Golden Willow. Particularly useful as ossier or basket-making willows.

Populus nigra, Black Poplar, thirty to forty feet, of rapid growth, wood valuable for flooring, cooperage, and for gunpowder charcoal.

Populus tremuloides, Mx. American Aspen, "Quaking Asp," twenty to fifty feet.

Populus nigra, var. *dilatata*, Lombardy Poplar, tall, spire-shaped tree, of rapid growth, to be set in rows for wind-breaks.

Populus monilifera, Ait. Necklace Poplar, large tree, one hundred and fifty feet high, light, soft wood, useful for box manufacture, and especially for paper pulp.

Juniperus Virginiana, L. North American Red Cedar, or Pencil Cedar. The largest of American junipers, sixty to ninety feet, furnishes a light, fragrant, and imperishable wood.

Abies concolor, Small Balsam Fir, or White Fir, eighty to one hundred feet.

Abies nobilis,
Abies magnifica, } Red Fir.

Abies religiosa, the Sacred Fir of Mexico.

" *bracteata,* one hundred feet.

" *Canadensis,* Mx. Common Hemlock, fifty to eighty feet.

Pseudotsuga Douglasii, Can. Douglas Spruce, one hundred and fifty to three hundred feet.

Picea Engelmanni, Eng. Spruce, sixty to one hundred feet.

" *pungens,* Eng. Balsam Spruce, sixty to one hundred feet.

Pinus flexilis, James. Pine, sixty feet.

Pinus ponderosa, Dougl. var. Scrophulorum, Eng. Yellow Pine, eighty to one hundred feet.

Salisburia adiantifolia (or Ginkgo biloba), forty to eighty feet.

Pinus edulis, Eng. The Piñon or Nut Pine, ten to fifteen feet.

Sequoya semperviceus, Sequoya gigantea, Red Wood, two hundred to three hundred feet, wood almost imperishable. Both these trees should be planted, as they are likely to become extinct unless rescued by cultivation.

Eucalyptus globulus, Blue Gum, peculiarly valuable in swampy or malarial districts. One hundred and fifty varieties of Eucalyptus, " the tree of the future." *Cf.* article in *Pop. Sci. Mo.,* vol. xii. : " The Eucalyptus of the Future," by Samuel Lockwood. See also, below, list by W. S. Lyon.

Eugenia jambos, Rose Apple, or Jamrosade, twenty to thirty feet, should be cultivated in the Southern States for its delicious fruit.

Catalpa bignonioides, Walt. Catalpa, thirty to fifty feet, a beautiful tree, possessing great advantages for timber, being the cheapest and easiest grown of all our forest trees, native or introduced, and also the most rapid in its growth.

Paulownia imperialis, Siebold. A grand flowering-tree, forty feet high, with immense leaves, of rapid growth, and particularly suited for parks, road-sides, and shade (from Japan), has large, purple, fragrant pannicles of flowers in the spring.

Schinus molle, the Pepper-Tree, or Peruvian Mastich-Tree. The leaves exude an oily fluid, filling the air with fragrance, particularly after a rain.

Eriobotrya japonica, the Loquat, or Mespilus, should be introduced into the Southern States, being one of the most grateful acid summer fruits. The tree is a very handsome, broad-leaved evergreen.

Cinnamomum camphora, the common Camphor-Tree of China. Growing trees can be had at the Agricultural Department, Washington, D. C.

Dryobalanopos aromatica, the Sumatra Camphor-Tree. Both of these should be introduced and cultivated in California.

Aleurites triloba, Candle-Nut-Tree, thirty feet, native of the Pacific islands, exceedingly useful for its oily nuts.

Ceratonia Siliqua, the Carob, or St. John's-Bread-Tree. Could easily be cultivated in immense numbers from seed, as is our ordinary locust-tree, and would be a valuable addition to the country, as it would furnish a large amount of food for cattle, the bread-bean pods being used for feeding cattle and swine in all countries where the trees grow, and is being imported largely into Europe and England.

Blighia sapida, the Akee. A native of West Africa, but becoming widely dispersed, would grow in the more Southern States, and furnishes a valuable fruit, very wholesome when cooked.

Persea gratissima, or Alvocada Pear, or Alligator Pear, a small tree, bearing large, purplish, pear-shaped fruits, much esteemed for dessert. Would be a valuable addition to the fruits of Southern California.

Prunus amygdalus, the Almond, fifteen feet. Already cultivated in California.

Carya olivæformis, N. Pecan Nut. Is proving one of the most valuable trees of Texas; recently introduced into Georgia; the yield of nuts is large, bringing good prices.

Castanea vesca, Spanish Chestnut. The tree furnishes a large percentage of the food of the poorer classes of Southern Europe, and its cultivation in this country should be encouraged. A Japanese "Giant" variety has been lately introduced. Said to be of better flavor than the Spanish chestnut.

Juglans regia, English Walnut, or Madeira Nut, sixty feet, valuable both for its wood and its nuts; the yield is large, as many as twenty-five thousand nuts to a tree.

Morus alba, White Mulberry, the most valuable for feeding silk-worms. Its cultivation should be encouraged. Several varieties are offered by the nursery-men for the large edible fruit.

Achras sapota, the Sapodilla, or Naseberry, a very sweet, high-flavored fruit. Tree spreading, with fine, glossy leaves.

Ægle marmelos, the Bael-fruit, Elephant Apple, Maredoo, or Bengal Quince, a small tree, of the orange family, producing an odd fruit and trifoliate leaves.

Anacardium occidentale, the Cashew-nut. A tree of the Terebinth family, attaining considerable size, and in growth resembling the walnut. The curious fruit is kidney-shaped, about an inch long, and after roasting is a good substitute for almonds, etc., at table.

Anona cherimolia, the Cherimoya, or Jamaica Apple, a loose, spreading tree of the Custon Apple family, attaining a height of twenty to twenty-five feet. The light-green fruit is beautiful, delicious, and considered one of the finest fruits of the world.

Anona muricata, Sour-Sop, fifteen to twenty feet high, fine glossy foliage, fruit large, heart-shaped (six to nine inches in circumference), green and prickly, contains a fresh, agreeable, sub-acid juice.

Chrysophyllum cainito, the Star Apple. A tree of thirty to forty feet, spreading branches, beautifully veined leaves, silvery white on the under-side, fruit about the size of an apple, wholesome, with an agreeable sweet flavor.

Ficus carica, Fig. Very easy of cultivation, and offered by the nursery-men in several varieties; should be largely cultivated in California and all our Southern States.

Malpighia glabra, Barbadoes Cherry. One of the favorite trees of the Barbadoes and West Indies, usually planted near dwellings, and as hedges. The trees are beautiful evergreens, bearing cherry-like fruit of a pleasant taste.

Mammea Americana, the Mammea Apple, or St. Domingo Apricot, sixty to seventy feet high, with broad, ovate, shining leaves; fruit angular, size of cocoanut, with juicy yellow pulp of delicious flavor.

Mangifera Indica, the Mango. This delightful fruit is now being introduced largely into Florida. It is of very rapid growth and fine form; five or six varieties are offered by nursery-men.

See "Popular Science," 1879, for an account of the ameliora-
tion of climate in dry barren districts by this tree.

Psidium Cattleyanum, the Cattley or Strawberry Guava, now
being much cultivated in Florida, is of fine appearance,
and the plum-like, claret-colored-fruit being of most
agreeable flavor.

Punica granatum, the Pomegranate. Easily grown and very
handsome; small trees, flowers showy. One of the most
desirable fruits. Bark of the tree used in medicine.

Psidium Guaiava, the ordinary Guava. One of the most valuable fruits for jellies and preserving. Several varieties
offered by nursery-men.

Tamarindus Indica, the Tamarind. A beautiful tree with delicate blossoms, and soft, pinnately divided leaves, grows to
eighty feet in height. The pods pressed in syrup or sugar
form the preserved tamarind of commerce.

Zizyphus jujuba, the Jujuba. A small tree of the Buckthorn
family, bearing small yellow, farinaceous, delicious berries.
The lotus spoken of by Pliny as furnishing the food of
the ancient Lybian people called Lotophagi.

Melia azedarach, L. The Bead-Tree, or Pride of India. Beautiful for streets and parks of our Southern cities; thirty to
forty feet high, flowers fine, loose, terminal, lilac-like spikes.

Dichopsis gutta. The Gutta-percha. This is a tree of the
Star-apple family, attaining a height of from sixty to
seventy feet. Leaves smooth, ovate, rusty-brown on under-side. This valuable tree is rapidly becoming extinct
in its native habitations, and efforts should be made to
introduce it into all tropical and subtropical climates.
The French government has recently decided to cultivate

it in Algeria. Why should it not be introduced into Southern California?

Mimusops globosa. The Ballata. Furnishes a milky juice equal to the best gutta-percha of the East. This should be tried. It is a native of British Guiana.

Butyrospermum Parkii, the Karite or Butter-Tree. An African tree, furnishing from its seeds the Shea-butter of commerce, used in soap-making, and a gum or coagulated juice which has recently been found to be equal to the best gutta-percha. It is possible that this tree might be made to grow in the warmer parts of California.

Cinchona calisaya, Cinchona succirubra, Cinchona condaminea. These three species of cinchona have been successfully cultivated in Mexico; in the Canton of Cordova several thousand cinchona-trees exist and are doing well. The beautiful trees with large velvety leaves, turning red when old, may be seen by those who travel by rail from Vera Cruz to Mexico. It is worthy of serious effort to cultivate this valuable tree in Texas and California.

Quillaja saponacia, Quillaia-bark Tree, Soap-bark Tree. This large tree (fifty to sixty feet) yields in its bark a product very valuable in cleaning delicate colored fabrics, and could undoubtedly be grown with profit in California.

The sub-tropical trees mentioned in the above list can all be obtained from the nursery-men (*e.g.,* Siebrecht & Wadley, Rose-Hill Nurseries, New Rochelle, New York), or of the Agricultural Bureau, Washington, D. C., and they can be successfully grown in any of the Southern States where the winter temperature does not fall below 45° F. I have not mentioned many well-known varieties, but have preferred to call attention to such little-known trees as might perhaps be cultivated (and, in fact, are already to some extent in Florida and California) successfully, with profit in various ways to our Southern States.

Very respectfully yours,
C. S. DOLLEY.

STATE BOARD OF FORESTRY,
LOS ANGELES, CALIFORNIA, June 6, 1892.

PROF. EDWIN J. HOUSTON,

    1809 Spring Garden St., Philadelphia, Pa.

DEAR SIR,—I send you a few "notes." The second part will deal with the much larger subject of lands *outside* of the natural forest districts, suitable *only* for forest planting, and the species we find most useful thereon. Our experiments with these are practicable and tangible, and I trust will prove of more value than the "glittering generalities" of the accompanying Part I.

I will endeavor within a week to give you the balance.

Very truly yours,
WM. S. LYON.

NOTES.

PART I.

ON TREES SUITABLE FOR REFORESTATION UPON THE PACIFIC COAST.

There are portions of the Sierra Nevada Mountains which, by the various processes of surface-mining, have been denuded of their original forest cover, as well as of every vestige of forest floor,—*i. e.*, fertile soil.

20*

In many localities the areas laid bare amount to hundreds of acres in single tracts.

In portions of Amador and Calaveras Counties, multitudes of "prospects" join each other, extending for miles upon a so-called "river-bar."

These "prospects" are frequently only holes that have been excavated from four to ten feet in depth, and the auriferous gravel thrown out and scattered far enough to nearly completely cover the original surface soil for from one to ten inches.

In Placer and Nevada Counties the process of hydraulic mining has generally washed out vast flats or valleys in the mountains, and leaving the resulting basin covered chiefly with gravel, boulders, or blue clay.

These operations have been discontinued in some localities for now more than thirty years; in others, only recently.

Yet in all of them, and apparently with an utter dearth of soil, the native timber is making an effort to assert itself.

At Dutch Flat, where gigantic mining operations have only ceased during three years, a very sparse setting of small *Psuedotsuga Douglasii, Pinus tuberculata, Libdocedrus decurrens,* and *Pinus Sabinana* have established themselves. The same features exist upon the Mokelumne River, in Calaveras County, where no mechanical disturbance has occurred since the original covering up of the soil thirty years before.

Both localities are between three thousand and three thousand six hundred feet in elevation, and the timber is such as belongs to that elevation in the Sierras.

The uninjured timber in the more southern county and adjoining denuded lands, can only be described as scrubby; and 1 cannot say that that which has established itself upon the worked-over sites has been notably depauperized by its mulch of clay and gravel. At the point of southern observation *Abies concolor* occurs, and seems to thrive.

I noted with interest that young plants of the Douglas spruce, growing about denuded fields in Nevada County, at three thousand five hundred feet, showed great vigor, although naturally fine specimens (mature) seldom occur in the primitive forest below five thousand feet.

It occurs to me that this is an index that this plant may prove valuable for future systematic reforestation. The other species named, except for purposes of forest cover, are not held in much esteem by lumbermen. At higher altitudes, and in the regions covering our valuable pines and spruces, extending over nearly seven hundred miles from San Bernardino to Plumas County, no mining operations (surface) have been conducted, and, except in isolated cases of torrential erosions, the forest floor is intact.

The original cover has been, however, heavily cut and burned over. Their native reproductive powers seem indestructible.

Where deforestation has been caused by the axe, a full proportion of young growths of *Pinus lambertiana, P. Coulteri, P. Jeffreyi. P. ponderosa,* and *Psuedotsuga Douglasii* seem to follow.

Where the denudation has occurred from fire, the more relatively worthless White Fir, Flat-leaved Cedar, and Hemlock seem to preponderate.

In the mountains,—*i. e.,* the natural timber district of the State,—not a single instance occurs where reforestation, upon any scale, large or small, has been undertaken.

No attempt or experiment with exotic species has ever been tried within the timber belt, and the values of the endemic species, and their readiness to conform to such unpromising situations as those described, justify us in thinking that they will best fulfil future systematic mountain reforestation.

PART II.

SPECIES USEFUL FOR REFORESTATION OF FOOT-HILL-LANDS, WASHES, OR LANDS UNSUITABLE FOR GENERAL AGRI-CULTURE IN CALIFORNIA.

(A) Endemic Species.

First among them I rate the *Pinus insignis,* or Monterey Pine. Naturally restricted to a narrow strip of land upon the Mid-California seaboard, not extending inland more than ten miles, it has taken kindly to transplantation to the interior hotter, drier valleys for more than one hundred miles from the sea. It rarely exceeds a height of thirty metres; the timber is generally twisted, coarse in grain, deficient in strength, not durable, and rates low for either lumber or fuel uses. Its pre-eminent value is as a forest cover and wide adaptability to soil and climate.

Extensive plantations upon arid, gravelly hill-sides, without care or cultivation, have attained an average height of twelve feet in five years from the seed, and isolated specimens have made a growth of sixteen feet during that time. It resists fire well, young plantations where badly burned over—*i.e.,* with three-quarters of the foliage destroyed—generally recuperating.

Cupressus macrocarpa, a smaller tree, fifty to seventy feet, nearly as local as the *Pinus insignis,* of still more rapid development, of more widely-tried distribution, and adapted to soils lacking in fertility; thrives wherever our rainfall reaches an average of sixteen inches, and resists extreme summer heat well; holds its lower branches and foliage with tenacity to a great age, making it serviceable for wind-breaks or ornamental hedging; the wood is light, but extremely durable underground, making it valuable for posts. Its small size makes it unavailable for general lumbering uses, other than as a cabinet wood, for which it is well adapted.

These two are the only native *conifers* that have been successfully planted upon lands deficient in moisture and fertility, upon a scale, and over a period of time prolonged enough, to assert that they will prove valuable for our so-called arid hill-sides.

In a more experimental way, upon more restricted areas, away from their native habitats, the subjoined coniferous species have been planted, and success has only followed where the soil was of reasonable depth, fair fertility, and where water was not remote from the surface. Such plantations have proven successful in the northern part of the State where water surface was distant, but only where the average annual rainfall exceeds twenty inches, and where the plantations have been made in nooks and valleys sheltered from drying winds.

In the southern half of the State, these conditions are not sufficient, and a deeply-cut, well-sheltered, cañon carrying water upon or near the surface is a *sine qua non*.

The species are :

Sequoia sempervirens.

Sequoia gigantea.

Chamæcyparis Lawsoniana.

All these are species reaching heroic dimensions, and the first and last rank in the very first class as unsurpassed timber trees.

The limits of their utility for forest planting is, owing to the conditions stated above, restricted to very narrow limits.

Pinus Sabiniana and *Juniperus California* both extend into the lower foot-hills, and naturally occur upon barren, stony lands of poorest quality ; and though nothing more than occasional experimental tests have been made, no question or doubt as to their utility upon utterly waste lands exists.

Their relatively *slower growth,* and inferior fuel and timber value, to the Monterey pine and Monterey cypress, explain their neglect, rather than any doubt as to their successful development.

For points where the conditions of soil and climate discourage the planting of the latter, these two species can be successfully introduced.

Of our native oaks, occasional tests have been made with a few species. The most valuable, the *Quercus densiflora*, makes but poor growth outside of humid and elevated ravines. *Quercus lobata* demands soil of both depth and fertility. Under these conditions it makes phenomenal growth. I have cut a tree displaying but forty-five annular rings, that measured four feet ten inches in diameter above any buttress.

Quercus agrifolia has proven most tractable of all upon dry, stony sites. Transplants easily, and after establishment makes fairly rapid growth. Its rather small size is an objection. The timber is inferior, stands but little transverse strain, but yields a superior fuel.

With our other trees no transplantations other than ornamental have been made of those extending below the coniferous timber belt.

Chief among them, the Oregon ash, Oregon maple, sycamore, cotton-wood, and laurel occur mostly in cañons, and with the exceptions noted are unsuitable for general planting.

The cotton-wood, *Populus Fremontii*, has shown (on a large scale) conformability to the lands strongly alkaline. Such are waste lands for this purpose, as the process of reclamation is tedious and costly. They comprise many thousands of acres of our West Coast bottoms, and no tree outside of this cotton-wood and *Tamarix gallica* (exotic) that has been tested upon them has heretofore proven satisfactory. Growth quick; timber warps badly; rates low for fuel.

The Laurel, *Umbellularia Californica*, has been attempted away from water-courses, with some measure of success. The timber is invaluable for veneers, exceedingly hard and heavy, and excels the cherry and redwood burl in beauty, but is of such exceedingly slow growth that few attempts have been made looking to its extended planting.

PART III.

(B) Exotic Species.

Considerable plantations have been made throughout the State with Eucalyptus species.

These in size range from one to four hundred acres. Plantations have been made with some fifty species, but ninety per cent. embrace,—

No. 1, *Eucalyptus globulus.*

Nine per cent. are of

No. 2, *Eucalyptus rostrata,* and

No. 3, *Eucalyptus virinalis*, and less than one per cent. of the other species.

No. 1, by reason of its more rapid growth, has been most freely planted. It is confined to the thermal belt, suffering when young from low temperatures (2° to 4° F. of frost). Nos. 2 and 3 are much hardier, and are *equally* resistant of drought.

All will grow upon arid hills, but only make remunerative growth where the subsoil is of an open, porous nature; the root *must* have an opportunity to get down. Upon rock or impervious subsoils these three gums may be seen, fairly vigorous, but not averaging over twelve to fifteen feet, now with stem diameter of over two and a half inches, representing five and six years' growth. Results about equal to one year upon open, porous soils of fair quality.

In value for fuel or timber, they take precedence as follows:

E. rostrata, 1.

" *globulus*, 2.

" *virinalis*, 3.

The former is reputed to be "teredo"-proof; but there is no timber as yet within the State large enough for wharf purposes.

In a smaller way, tests have been made with other species; the following prove of greatest merit:

Euc. corynocalyx (Sugar Gum). Hard and durable timber. Tree umbrageous, endures drought and sterile soils. Susceptible to light frosts.

Euc. diversicolor, characteristics not dissimilar from last.

Euc. populi folia, smaller tree than last, but hardier.

Euc. leucoxylon, rose-flowered variety, and

Euc. amygdalina, var. *angustifolia,* are the only two species that extend beyond our extra-tropical limits, having successfully withstood the cold of 16° F.

Have not seen the *E. amygdalina* tested upon poor soils. The former, however, is very tenacious of life upon dry and poor lands.

Both make trees of the greatest magnitude.

In cultivable lands, *Euc. Gunni* will make more rapid growth than *E. globulus,* and hence is profitable for a fuel crop that can be cut every four years.

A few other promising sorts are,—

E. citriodora, for its essential oil.   Dry exposures.

E. paniculata.   Dry exposures.

E. punctata.   Dry exposures.

Most species of *Eucalypti* make good fuel; a few are very durable under ground or in water, and though largely used in the antipodes for railway building, ties, etc., are in disrepute here from the tendency of some species to check and warp, disabilities that can be overcome by cutting in proper season and reasonable attempts at curing.

Acacias.

Upon any soil non-alkaline, and wherever the rainfall approximates sixteen inches, these are indicated for general forest uses.

They require but little assistance to be established, attain marketable size in five to eight years, furnish an excellent fuel (wood too small for most economic purposes), and some excel all other trees in the quantity and quality of superior tanbark yielded to the acre.   They have been fairly tested upon hill, valley, and brush lands, where cultivation was impracticable; the results have been promising, although, like the eucalypti, a measurable and perhaps profitable increment in growth has followed the cost of cultivation.

The most valuable species—*i.e.*, the one richest in tannin—is the *Acacia pycnantha.* It is, however, more sensitive to cold than some others, and hence properly restricted to the thermal section of the coast.

The next in value. *A. decurrens,* and its immediate congener, *A. moligsima,* are suitable for a very wide range, thriving in the littoral regions, as well as for one hundred miles inland, and for a length of seven hundred miles north and south.

A. melanoxylon is less valuable for its tan-bark, but makes a larger tree (twenty metres), and furnishes a valuable cooper's wood and the best fuel. It requires, perhaps, more moisture and a better soil than the other species to obtain its maximum growth. Incidentally it is compact and symmetrical in habit, hence serving well for wind-breaks or street-planting.

Casuarinas of different sorts have been fairly tried. Among them *C. stricta, C. teniusimus, C. suberosa,* and *C. equisitacetolia.* All are rapid growers and hardy, also adapted to arid sites.

The last-named has given larger evidence of versatile adaptability to our requirements of soil and climate.

Allied to the acacias we have a tree of phenomenally rapid growth, and conformable to waste lands. It is *Albizzia lophantha.* It is of no possible economic value except for the marvellous rapidity with which it furnishes forest cover, and is thus rendered available for furnishing a quick and short-lived protection to coniferous plantations.

Eastern United States Silva have been but sparingly attempted. Reasonable success has followed the planting of both *Catalpa* and *Robina psuedo Acacia* in the northern half of the State. In the south they exigently demand soil of greater depth and fertility than we are willing to classify as forest lands. The same holds true of most of the nut-bearers, the *Caryas* and allies, although sporadic and unprofitable attempts have been made with the English, the Black Walnut, and the Pistachio.

Some plantations of *Pinus pinaster* give promise of doing well, remote from the seaboard, and, to a limited extent, *P. Austriaca, P. strobus, P. cembra,* and *P. Laricio* have been planted.

Results, as far as obtained, indicate unsatisfactory growth, although the limited period of observation makes it premature to formulate any conclusions.

Isolated cases exist of above fifty additional exotic species that have been planted for forest uses.

The data is sufficient to furnish material that is not largely hypothetical.

If such be desired, a synoptical list can be supplied.

WILLIAM S. LYON.

INDEX.

Forest, droughts produced by destruction of, 81.

encroachments on, necessity for, 11.

influence of avalanches in destruction of, 67.

influence of inundations in destruction of, 66.

influence of, on humidity of air, 97.

influence of, on natural drainage, 113.

influence of, on preventing disastrous frosts, 130, 131.

influence of, on rapidity of evaporation, 97.

influence of parasitic plants on destruction of, 71, 72.

influence of rodents on destruction of, 73.

influence of torrents on destruction of, 58.

influence of wind on destruction of, 66.

inundations produced by destruction of, 81.

loss of soil following destruction of, 81.

malarious diseases produced by destruction of, 82.

natural drainage disturbed by destruction of, 82.

necessity for preservation of exact balance of conditions to existence of, 63.

necessity for removal of, 11.

of mountain slopes, 46.

influence of, on rapidity of drainage, 104.

Forest of tropical regions, 46.

peculiar distribution of moisture causing, 43.

products of, 11.

protection afforded by, against frosts, 129.

why a regular distribution of moisture is requisite for growth of, 44, 45.

why the air of, is cooler and damper in summer than air over open fields in same district, 132.

Formation of soil, 50 to 57.

Freezing and melting of ice, influence of, on disintegration of rocks, 52, 53.

G.

Gases, absorption of, by soils, 56.

Geikie, Archibald, extract from "Text-Book of Geology," 60, 78, 88, 133.

Germ, necessity for existence of, for plant life, 23.

Germ-cell, protoplasm of, 20.

Germs of plants, source of, 21.

wonderful vitality of, 21, 33, 34, 35, 36.

of seeds, wonderful vitality of, 33, 34, 35, 36.

Glaciers, definition of, 54

influence of, in formation of soil, 54, 59.

Goats, destruction of the forests by, 73.

Grasshoppers, destruction of forests by, 73, 74.

Gravelly soils, definition of, 50.

O.

Oceanic climate, peculiarities of, 123.
Oxygen of atmosphere, use of, 140.
   necessity for, in respiration of animals, 141.

P.

Paragrêles, 150.
Parasitic plants, influence of, in destruction of forests, 71, 72.
Peaty soils, definition of, 51.
Pine forests, appearance of scrub oak on burning over of, 36, 37.
Plant germs, nature's method of distributing, 32.
   wide distribution of, 32 to 38.
   wonderful vitality of, 33, 34, 35, 36.
   nationality, 26.
   seeds, wonderful vitality of, 33, 34, 35, 36.
Planting of trees, when advisable, 157.
Plants, animals, and minerals, mutual interdependence between, 168.
   conditions necessary for the growth of, 20 to 24.
   decaying, influence of, in formation of soil, 50.
   necessity of oxygen for growth of, 23.
   source of germs of, 21.
   source of seeds of, 21.
Polar currents of air, why generally drought-producing, 101.
   land areas, effect of, on earth's climate, 167.

Prairies, new growth of plants marking wagon-tracks on, 36.
Precipitation, forms of, 100.
Pressure of air, influence of, on rapidity of evaporation, 92.
Prestwich, Joseph, extract from "Geology, Chemical, Physical, and Stratigraphical," 117, 118.
Products of the forests, 11.
Protoids, definition of, 29.
Protoplasm of germ-cell, 20.
Pouchet, extract from "The Universe," 78, 79, 80.
Purification of the atmosphere, 140 to 145.

R.

Railway sleepers, demands on forests for, 89.
Rain, 100.
   causes of, 100.
   Huxley on the distribution of, 106.
   Maury on the distribution of, 107, 108.
Rainfall, distribution of, 102.
   true index of the wealth of a country, 94.
Reclus, Élisée, extract from "Earth," 67, 68, 69, 116, 117.
   extract from "Ocean," 30, 31, 48, 49.
Red River, raft in, 67.
Reforestation, 155 to 159.
   and deforestation, 14.
   definition of, 155.
   necessity for government encouragement of, 156.
   objects of, 155, 156.

THE END.

www.ingramcontent.com/pod-product-compliance
Lightning Source LLC
Chambersburg PA
CBHW020850270326
41928CB00006B/634